THE BIODYNAMIC SOWING AND PLANTING CALENDAR

2005

The original biodynamic sowing and planting calendar showing the optimum days for sowing, pruning and harvesting various plant crops, as well as for beekeeping

Compiled and edited by Maria and Matthias Thun

Floris Books

Translated by Bernard Jarman

Published in German under the title *Aussaattage.*
English edition published by Floris Books

British Library CIP Data available

ISBN 0-86315-458-1
ISSN: 1478-1573

Printed in Poland

Walter Thun, Swans on the River Lahn, *Öil*

Swans on the River Lahn

On the River Lahn in the district of Gisselberg not far from Marburg, a certain Andreas Ortwein made some land available to us. Here we were able to continue with the trials we had started in Marburg. Often at dawn while stirring and spraying out the horn silica preparation, we often had powerful experience of the living-etheric world. The elemental spirits were tangibly close. It is wonderful how our spray preparations can have such a positive effect on plant growth.

The artist Walter Thun regularly came to this place in search of new motifs for his work. His painting *Swans on the River Lahn* recalls this Light-filled mood.

Looking back over 2004

The first three months were very changeable having both cold periods and spells with summer-like temperatures. Whenever Warmth trines occurred temperatures rose to at least 20°C. Already in early February during what is normally the European winter, there were flowering spring landscapes. By contrast the first week of April saw numerous car accidents caused by drifting snow and over Easter night time temperatures fell to –4°C. Although partly sunny during the day a cold north-east wind blew. We had predicted such weather in the calendar of 2004.

With great excitement we awaited the rare transit of Venus across the Sun's disc on June 7, 2004. Venus had reached its node the day before when it was possible to experience unusually strong Light and Warmth effects. This runs counter to what normally occurs when Venus is at its descending node. With Mercury for example the descending node always brings cooler weather.

On this occasion we experienced the first warm nights of the year. The day dawned clear and cloudless. Shortly after sunrise we were able to see, using the telescope, two dark areas on the Sun. These were probably sunspots since they didn't move. We had often looked at them. Then at the alotted time Venus appeared as a black spot moving across the face of the Sun. The temperature was steadily rising and it soon became almost too hot.

Above: Photographing the transit of Venus with Supertele and Telescope.

Top left: Venus entering the sun's disk.

Left: Venus leaving the sun's disk.

We had of course planned some sowing trials. These began on the evening of June 6. From then on we sowed seeds every two hours right through to the morning of June 9. Matthias prepared the soil on the Heckacker research plot and sowed radishes and mustard. At the same time I sowed spinach in the greenhouse. I also set some mustard seed to germinate in a petrie dish. The hot weather continued until June 9, then it reverted to the previous conditions and a cold east wind started blowing again.

Sunrise June 6, 2004

Sowing mustard and radishes

Mustard and radishes growing

Mustard sprouts in petrie dishes

Spinach sprouts

Prognosis for 2005

It is usually assumed that varying weather phenomena are connected with changes taking place in the environment. The following is also worth bearing in mind however. For forty years it is continually confirmed to me that the vagaries of the weather are connected to a considerable degree with the rhythmic movements of the planets. Consider for instance the trine positions over three years.

In 2003 there were 10 Water, 7 Earth, 14 Light und 2 Warmth trines.
In 2004 there were 7 Water, 8 Earth, 4 Light und 6 Warmth trines.
In 2005 there will be 7 Water, 11 Earth, 9 Light und 2 Warmth trines.

We can also consider the planetary oppositions.

In 2003 there were 28 Light–Warmth oppositions and 15 Earth–Water ones.
In 2004 there were 6 Light–Warmth oppositions and 9 Earth–Water ones.
In 2005 there will be 9 Light–Warmth oppositions and 15 Earth–Water ones.

If the position of each planet against the background of the constellations is considered along with their periods of retrograde movement — when they have their greatest effect on the Earth — then it becomes transparently clear that weather conditions are influenced by the rhythmic movements of the planets. Thus periods of flooding are usually linked to a Water trine.

Returning to our considerations for 2005, this year Mercury has its greatest influence during its retrograde phases when standing in Watery constellations. At the beginning of the year Venus is also in front of the Watery constellation of Scorpion. She then moves with and gradually overtakes the Sun to begin her retrograde movement around Christmas time in the Goat. The time of her greatest influence therefore in January 2006.

In January Mars also moves through the Watery constellation of the Scorpion and then limps slowly after the Sun to find himself during October in the constellation of the Ram where he turns retrograde. At the end of last year Jupiter entered the Virgin. Here he will remain and cause no great excitement. Saturn starts the year in front of the Twins and later in the summer moves into the Crab where his influence is relatively weak. Uranus remains in the region of Waterman while Neptune is in the winter constellation of the Goat and Pluto in the Watery Scorpion.

Eclipses generally have a negative influence on the course of the year. There are two solar and two lunar eclipses this year as there are in most years. Jupiter will be affected particularly strongly having nine occultations by the Moon. In addition there will be four occultations of Venus, two of Mars and two of Mercury. Oppositions to the Moon occurring during these periods will cause disturbances in the earth. As a result many days have had to be marked as unfavourable in the calendar.

Saturn at its ascending node in January

During each rotation all the planets cross the ecliptic two times, once in an ascending (☊) and once in a descending (☋) direction. The Moon has a negative effect when sowings are made at these times and therefore three to four hours are marked as unfavourable in the calendar. Mercury is at a node on about seven days during the year. Saturn by contrast has two nodal periods during the thirty years it takes to complete one rotation. We have observed that it takes about four and a half months for the nodal effect to be overcome. Many insects have a strong relationship to Saturn and appear to thrive when it is at the ascending node. The planet is on its ascending node on January 8. This means that the nodal effect will already be felt at the end of October 2004 and hence a reduction in the vitality of seeds sown during the Moon-Saturn conjunction can be expected. These periods are therefore also marked as unfavourable non-sowing days.

The last time that Saturn was at its ascending node was thirty years ago, during August 1975. In that year huge numbers of Flame Shoulder moths *(Ochropleura plecta)* appeared and became a major pest. On that occasion they were particularly partial to carrots and destroyed acres of them in a very short span of time. We were then working together with Professor von Boguslawki on some large scale carrot trials at Rauischholzhausen. These plants were also not spared the depredations of the Flame Shoulder caterpillars. Each morning when I went to the carrot fields another four or five metres had been devoured. The roots were eaten clean off and only the green tops remained. This was repeated every night. The whole matter was however somewhat mysterious. Three rows were always eaten off and three rows remained standing. Sometimes all six rows were eaten. This curious threefoldness was a complete riddle to us.

Some days later when I once again investigated the matter I discovered that the rows of carrots which had been eaten off all had green tops. The others did not. I then went to observe the women who did the hoeing. I saw how on some rows they were hoeing away from the plants while on others towards them, covering the roots with earth. I then realized that if the top of the carrot is exposed it develops chlorophyl and that this provides the ideal nourishment for the Flame Shoulder larvae. When laying her eggs the adult moth knew exactly which roots contained the chlorophyl.

The Flame Shoulder caterpillars did not of course only appear on biodynamic farms but also where mineral fertilizers were applied. We asked around to find out what pesticide was being used to combat them and discovered that none existed because for the last fifteen years this caterpillar had not been a pest! Now fifteen years ago Saturn had also been at its nodal point and the damage for the grower was a favourable situation for the insect!

From all this we learnt the value of lightly earthing up the carrots and this also improved their quality. This time the effect of Saturn's node will be felt from the end of October 2004 to about the middle of March 2005. Which specific insect will profit from this we do not yet know, nor whether similar pests in other parts of the world will be encouraged. We would be extremely grateful to receive information about this.

Some fundamentals of plant production

The seeds of wild plants can spread far and wide. With the help of the wind seeds are transported over vast distances and even across continents before being deposited. If soil conditions are favourable plants will grow, thrive and make themselves at home. Where growing conditions are not suited they will soon be choked out by others.

Only a few of these wild plants are used for human food. More than seven thousand years ago the great leaders of humankind developed insights into the life of plants and the creative powers of the cosmos which directed them. Soil cultivation began and food plants were developed out of an understanding for the cosmic laws of Sun, Moon, planets and even the fixed stars. In this way our cultivated plants came into existence. Today we can look back over a six- to seven-thousand-year history of cereal, vegetable, berry and fruit production.

In addition to food plants the oriental courts were keen on developing flavours and scents. Gradually a rich culture of herbs and spices evolved. Not only were leaves, flowers and fruits used but also the special flavours found in seeds, barks and roots. These helped to improve the taste and digestibility of the main food crops. During the course of the millennia, a whole new group of wild plants also developed on the tilled lands and began to compete with the crops — the so-called weeds. Since most of them had medicinal properties, their root excretions were able to bring qualitative improvements to the planted crops. This observation gradually led to the search for substances that would benefit plant growth and ultimately to the ground-breaking work of Justus von Liebig.

We must learn to see the soil as a living organism and take care of its humus content to ensure that it suffers neither from nutrient excesses nor deficiencies. Both extremes can have a negative effect on crop health.

If a one-sided cropping system or monoculture is followed year after year, the plants will become increasingly susceptible to disease and pests will appear. We can take the cabbage or brassica family as an example. Continual one-sided cropping leads to disturbances in the calcium processes of the soil and club-root develops. It is a fungal infection that causes growths to form on the roots. Spores of the fungus remain active in the soil for four years. If during this time any kind of brassica, whether cabbage, radish or rape, is planted the fungus will develop anew and become an ongoing problem. To address this one fungus problem, a five-year crop rotation is necessary.

If unrotted organic materials are added to soil in which carrots, onions or leeks are to grow, the corresponding root flies will soon arrive and lay their eggs because then the ideal conditions will have been provided for their maggots to thrive.

It is good practice to make a careful drawing of the cropping layout each year because with the diversity of crops being planted it becomes hard to remember what has been planted on the various beds three years down the line.

A lot is known today about the specific nutrient needs of cultivated plants. It can be read on every seed packet. We know very little however about what the plants leave behind in the soil or what the varieties we sow or plant meet in the soil as an inheritance from their predecessors. This is why it is important to compost these areas in the autumn. In addition to having a good crop rotation the vegetable gardener needs to understand how to use well ripened earthy compost in the right way.

Well ripened earthy compost

Pure plant-based compost should be made in order to harmonize the vegetable plot. There should also be compost that has been enriched with animal manure or animal remains. Heavy feeders like cauliflower or potatoes need this animal element in order to develop.

If substances of animal origin are used such as hoof, horn and bone meal, it is vital that they are thoroughly composted beforehand together with vegetable matter. These and other animal remains need to be composted with plant material because animal forces are still present. If they are added directly and have to be broken down in the soil, a fungal process takes place that can be transferred to the growing plants. This is why the breakdown process needs to be completed in the compost heap beforehand.

Biodynamic gardeners use compost preparations to help transform the organic material. Various companies offer compost activators to speed up the composting process. Our cow pat preparation (or manure concentrate) can also be used for this purpose by spraying it now and again during the year over the compost pile. How to make it is described in detail in our book *Gardening for Life.* Because of the large amount of well formed cow manure that is needed, only farmers or market gardeners will be in a position to make it themselves. Domestic gardeners can try to obtain the preparation ready made. We also run courses on making compost at our research centre from time to time.

Pure plant-based compost can be spread on all the beds while that which incorporates animal materials or manures can only be put where plants will be grown that love manure.

Compost heaps

The zodiac

The **zodiac** is a group of twelve constellations of stars which the Sun, Moon and all the planets pass on their circuits. The Sun's annual path always takes exactly the same line called **ecliptic.** The Moon's and planet's paths vary slightly, sometimes above and sometimes below the ecliptic. The point at which their paths cross the ecliptic is called a **node.**

If New Moon is at a node there is a **solar eclipse,** as the Moon is directly in front of the Sun, while Full Moon at a node causes a **lunar eclipse** where the Earth's shadow falls on the Moon. If Sun or Moon pass exactly in front of a planet there is an **occultation.** If Mercury or Venus pass exactly in front of the Sun, this is a **transit** (other planets cannot pass in front of the Sun).

The angles between Sun, Moon and planets are called **aspects.** In this calendar the most important is the 120° angle, or **trine.**

In this illustration the outer circle shows the *varying* sizes of the visible **constellations** of the zodiac. The dates on this outer circle are the days on which the Sun enters the constellation (this can change by one day because of leap years). The inner circle shows the divisions into *equal* sections of 30° corresponding to the **signs** used in astrology.

It is the *constellations* on which Maria Thun's observations are based, and which are used throughout this calendar. Both Latin and English names are used indiscriminately.

Trines

The twelve constellations are grouped into four different types, which each have three constellations at an angle of about 120°, or trine. About every nine days the Moon passes a similar region of forces. Through hoeing or spraying silica (501) in the rhythm of the trines we can reawaken the impulse of the original sowing day.

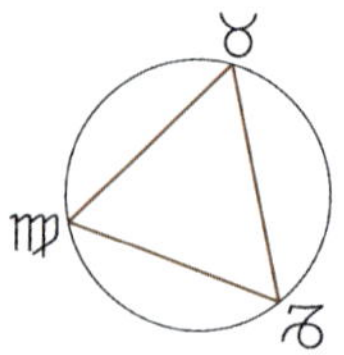

Earth-Root

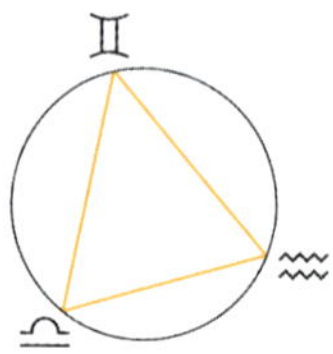

Light-Flower

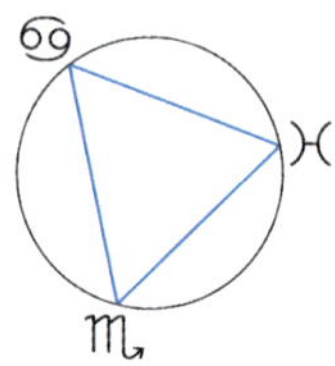

Water-Leaf

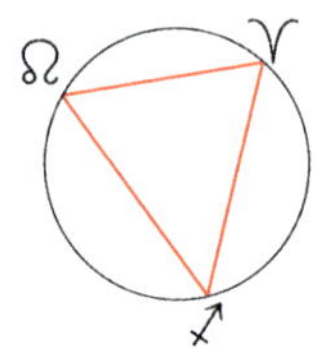

Warmth-Fruit/Seed

The sidereal Moon

In its 27-day orbit round the Earth the Moon passes the twelve regions of the zodiac and transmits forces to the Earth which find expression in the four elements. They bring about enlivening processes in the four different plant organs. At the point of sowing, cultivation and harvest we can stimulate the growth and health of a plant.

These cosmic forces work similarly in bees and can be used to enhance their life. The colony shuts itself away from the outer world in the skep or box by sealing it with propolis. When we open the skep it causes a certain 'confusion' in the colony. A new cosmic impulse can work into this unrest, directing the activity of the bees until the next routine disruption.

Let us summarize the laws pertaining to plant-experiment, beekeeping and weather observation in one table (opposite).

The different impulses vary between two and four days. This basic framework is sometimes interrupted. It can happen that planetary oppositions override some days with their different impulses. It may be also that trine positions activate a different element to the one the Moon is transmitting on this day. Days when either the Moon's path intersects with the ecliptic (ascending ☊ or descending ☋ node) produce mainly negative effects which are intensified if there is an eclipse or occultation during which the influence of the more distant planet is interrupted or changed by the nearer one. Such times are unsuitable for sowing and harvesting.

Constellation	*Sign*	*Element*	*Weather*	*Plant*	*Bees*
Pisces, Fishes	♓	Water	Watery	Leaf	Making honey
Aries, Ram	♈	Warmth	Warm	Fruit	Gathering nectar
Taurus, Bull	♉	Earth	Cool/cold	Root	Building comb
Gemini, Twins	♊	Light	Airy/bright	Flower	Gathering pollen
Cancer, Crab	♋	Water	Watery	Leaf	Making honey
Leo, Lion	♌	Warmth	Warm	Fruit	Gathering nectar
Virgo, Virgin	♍	Earth	Cool/cold	Root	Building comb
Libra, Scales	♎	Light	Airy/bright	Flower	Gathering pollen
Scorpio, Scorpion	♏	Water	Watery	Leaf	Making honey
Sagittarius, Archer	♐	Warmth	Warm	Fruit	Gathering nectar
Capricorn, Goat	♑	Earth	Cool/cold	Root	Building comb
Aquarius, Waterman	♒	Light	Airy/bright	Flower	Gathering pollen

Groupings of plants for sowing and harvesting

In cultivating a plant, particular parts are developed for food. We can divide them into four groups.

Root crops on Root Days

Development in the root region is in radishes, swedes, sugar beet, beetroot celeriac, carrot, scorzonera, etc. Potatoes and onions are included in this group too. These days produce good yields and top storage quality.

Leaf plants on Leaf Days

Development in the leaf realm is in the cabbage family, lettuce spinach, lambs lettuce, endive, parsley, leafy herbs and fodder plants. Leaf days are suitable for sowing and tending these plants but not for harvesting and storage. For these (as well as harvesting of cabbage for sauerkraut) Fruit and Flower days are recommended.

Flower plants on Flower Days

These days are favourable for sowing and tending all kinds of flower plants but also for cultivating and spraying 501 on oil-bearing plants such as linseed, rape, sunflower, etc. Cut flowers have the strongest scent and remain fresh for longer if cut on Flower days, and the mother plant brings forth many new side shoots. If flowers for drying are harvested on Flower days they retain the most vivid colours. If cut on other days they soon lose their colour. Oil-bearing plants are best harvested on Flower days.

Fruit Plants on Fruit Days

To this category belong all those plants which have their 'fruit' in the realm of the seed such as beans, peas, lentils, soya, maize, tomatoes, cucumber, pumpkin, courgettes, but also cereals for summer and winter crops. Sowing oil-bearing plants brings the best yields of seeds. The best time for extraction of oil later on is on Flower days. To grow good seed, Leo days are particularly suitable. Fruit plants are best harvested on Fruit days. They promote a good quality of storage and support regeneration. If storing fruit additionally choose the time of the ascending Moon.

There is always some uncertainty as to which category some plants belong. Onions and beetroot give a somewhat similar yield when sown on Root and Leaf days, but the keeping quality is best from Root days. Kohlrabi and cauliflowers belong to Leaf days, as does Florence fennel. Broccoli is more beautiful and firmer when sown on Flower days.

Explanations of the calendar pages

Next to the date is the constellation (and time of entry) in which the Moon is. This is the astronomical constellation, not the astrological sign, see page 11. In the next column astronomical aspects are indicated which have relevance for plant growth (see Astronomical Symbols).

A further column shows which element works predominantly on that day (this is useful for beekeepers). Sometimes there is change during the day when both elements are mentioned in order. Warmth effects on thundery days are implied but are not mentioned in this column, but may have a ♄ symbol in the far right.

The next column shows in colour the part of the plant which will be enhanced by sowing or cultivation on that day. Numbers indicate times of day. On the extreme right special events in nature are noted as well as anticipated weather changes which disturb or break up the overall weather pattern.

When elements are indicated which do not correspond to the Moon's position in the zodiac (sometimes it may even be more than one element on the same day), it is not a misprint, but takes account of other cosmic aspects which overrule the Moon-zodiac pattern and have an effect on a different part of the plant.

Unfavourable times are marked thus (----------). These are caused by eclipses, nodal points of the Moon or the planets or other aspects with a negative influence;

they are not elaborated in the Calendar. If one has to sow at unfavourable times for practical reasons, one can choose favourable days for hoeing and so bring about an improvement.

Astronomical symbols

Other symbols

Constellations		*Planets*		*Aspects*			
♓	Pisces	☉	Sun	☊	ascending node	**St**	Storms likely
♈	Aries	⊕	Earth	☋	descending node	ϟ	Thunder likely
♉	Taurus	♀	Venus	☊	Highest Moon	**Eq**	Earthquakes
♊	Gemini	☿	Mercury	☋	Lowest Moon	**Tr**	Traffic dangers
♋	Cancer	♂	Mars	**Pg**	Perigee	**Vo**	Volcanic activity
♌	Leo	♃	Jupiter	**Ag**	Apogee	**UW**	Unsettled weather
♍	Virgo	♄	Saturn	☍	Opposition	**H**	Hail likely
♎	Libra	♅	Uranus	☌	Conjunction		
♏	Scorpio	♆	Neptune	•	Eclipse, occult'n		Northern Planting Time
♐	Sagittarius	♇	Pluto	••	Lunar eclipse		
♑	Capricorn	○	Full Moon	△	Trine		Southern Planting Time
♒	Aquarius	●	New Moon				

Two-Weekly Planting Times

The ascending and descending Moon

After the Moon during its monthly orbit reaches its lowest position in the zodiac, in front of the Archer, it begins to ascend again. Its daily arcs across the sky become larger every day. Its rising point shifts north-eastwards and the setting point moves towards north-west. This time of **ascending Moon** should not be confused with the waxing Moon.

During the time of the ascending Moon plant sap rises more strongly. The upper parts of the plant become filled with sap and vitality. It is a favourable time for cutting scions (for grafting). The effect can be enhanced by choosing Fruit days within this period for fruit-bearing plants, and Flower days for flowering plants. The same applies for the times of grafting. Fruit harvested during this period remains fresh for a longer time when stored. This time is also suitable for cutting Christmas trees. They keep their needles longer. If cut on Flower days their fragrance is most pleasant.

The ascending Moon

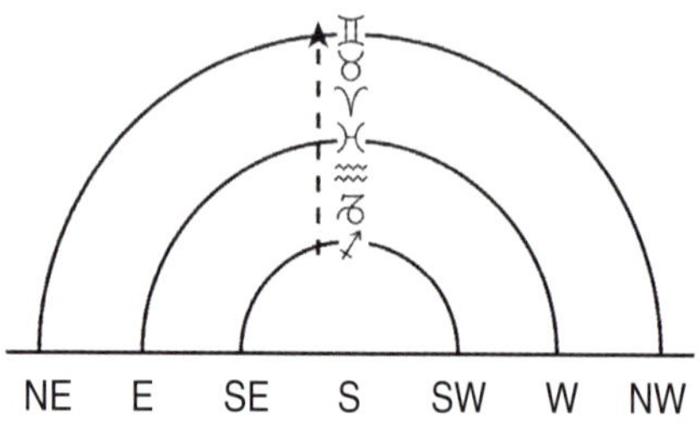

The descending Moon: Planting Time

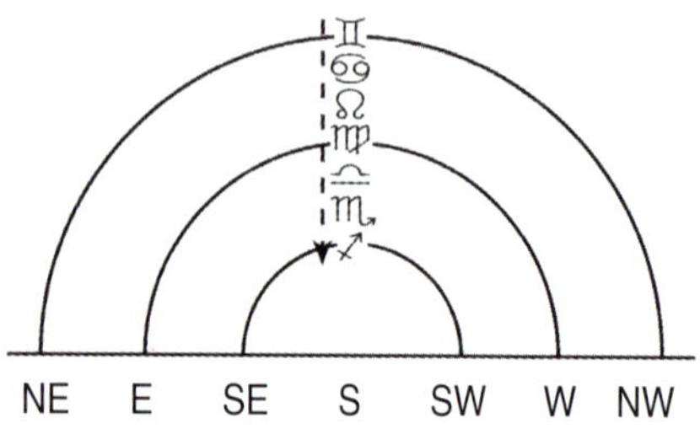

After the Moon reaches its highest point, in the Twins, it becomes **descending.** Its arcs in the sky become lower every day, its rising point shifting towards south-east and its setting towards south-west.

By sowing we mean when the seed is first put into the soil. Moving plants from one position to another is transplanting, for instance when we move young plants from their seed bed to their final beds. For both these operations choose the period of the descending Moon (not to be confused with the waning Moon).

During this Planting Time, plants take root readily and thus connect quickly with their new location. This impulse can be further strengthened for the particular type of plant, if one chooses for leaf plants the corresponding Leaf days during this planting period (Moon in Crab or in Scorpion). For root-crops such as celeriac, choose any of the Root days (Moon in the Virgin) or for fruiting vegetables, such as cucumber or tomato, choose the corresponding Fruit days (Moon in the Lion). In this way an enhancement of the particular development can be achieved, in addition to the rooting impulse.

During this period the sap rises only feebly. That is why at these times hedge trimming, tree pruning, timber felling as well as manuring of meadows, pastures and orchards can be recommended.

The southern hemisphere

During the Moon's movement from Archer (high in the northern sky) to Gemini (low) the Moon is descending, the rising and setting points moving northwards. Conversely, during the northern hemisphere's descending time, in the southern hemisphere the Moon is ascending.

For this reason there are separate (and opposite) planting times given in the calendar. Northern Planting Times refer to the northern hemisphere, and Southern Planting Times refer to the southern hemisphere. All other constellations and planetary aspects are equally valid in both hemispheres.

Times

Times given are *Greenwich Mean Time,* using 24-hour clock with ^{h} after the time. Thus 15^h is 3 pm. ***No account is taken of daylight saving (summer) time (DST)***.

For different countries adjust as follows:

Britain, Ireland, Portugal, Iceland: GMT (DST from March 27 to Oct 29, add 1^h)
Central Europe: add 1^h (DST from March 27 to Oct 29, add 2^h)
Eastern Europe (Finland, Baltic states, etc.): add 2^h (DST from March 27 to Oct 29, add 3^h)

Namibia: add 1^h (DST to April 2 and from Sep 3, add 2^h)
Israel: add 2^h (DST from probably April 1 to Sep 1, add 3^h)
Egypt: add 2^h (DST from April 29 to Sep 29, add 3^h)
South Africa: add 2^h (no DST)
Kenya: add 3^h (no DST)
Pakistan: add 5^h (no DST)
India: add $5^1/_2{}^h$ (no DST)
Bangladesh: add 6^h (no DST)

Western Australia: add 8^h (no DST)
Northern Territory: add $9^1/_2{}^h$ (no DST)
South Australia: add $9^1/_2{}^h$ (DST to March 26 and from Oct 30, add $10^1/_2{}^h$)
Queensland: add 10^h (no DST)
ACT, NSW, Victoria: add 10^h (DST to March 26 and from Oct 30, add 11^h)
Tasmania: add 10^h (DST to March 26 and from Oct 2, add 11^h)
New Zealand: add 12^h (DST to March 19 and from Oct 2, add 13^h)

Brazil (Eastern): subtract 3^h (DST to Feb 26 and from Oct 2, subtract 2^h)
Newfoundland Standard Time: subtract $3^1/_2{}^h$ (DST April 3–Oct 29, subtract $2^1/_2{}^h$)
Atlantic Standard Time: subtract 4^h (DST April 3–Oct 29, subtract 3^h)
Eastern Standard Time: subtract 5^h (DST, except *IN,* April 3–Oct 29, subtract 4^h)
Central Standard Time: subtract 6^h (DST, except *Sask.* April 3–Oct 29, subtract 5^h)
Mountain Standard Time: subtract 7^h (DST, except *AZ,* April 3–Oct 29, subtract 6^h)
Pacific Standard Time: subtract 8^h (DST April 3–Oct 29, subtract 7^h)
Alaska Standard Time: subtract 9^h (DST April 3–Oct 29, subtract 8^h)
Hawaii Standard Time: subtract 10^h (no DST)

January 2005

Date	Const. of Moon	Other aspects	Moon Element		Parts of the plant enhanced by Moon or planets	Weather, etc.
1 Sat	♌	☉-♐	Warmth	NPT	Fruit	St
2 Sun	♍ 3^h		Warmth/Earth	NPT	Fruit to 2^h, Root from 3	
3 Mon	♍		Earth	NPT	Root to 20^h	Tr
4 Tue	♍	☋22^h	Earth	NPT	Root from 4^h to 20^h	Tr
5 Wed	♎ 16^h		Earth/Light	NPT	Root from 2^h to 15^h, Flower from 16^h	Eq
6 Thu	♎		Light		Flower to 11^h ----------------	
7 Fri	♏		Water		--------------------------------	
8 Sat	♏	♄☊	Water		--------------------------------	St E
9 Sun	♐ 1^h	◡11^h	Warmth		Fruit from 1^h to 19^h	ϟ Eq Tr
10 Mon	♐	●12^h **Pg** 10^h	Warmth		---------------- Fruit from 19^h to 23^h	Tr
11 Tue	♑ 0^h		Earth	Southern Planting Time	Root from 0^h to 23^h	
12 Wed	♒ 21^h	☿☋19^h	Earth/Light	Southern Planting Time	--------------------------------	
13 Thu	♒		Light	Southern Planting Time	--------------------------------	
14 Fri	♓ 14^h	☉☍♄	Light/Water	Southern Planting Time	--------------------------------	
15 Sat	♓	♂☋2^h	Water	Southern Planting Time	--------------------------------	St
16 Sun	♓		Water	Southern Planting Time	Leaf from 0^h to 9^h, and from 15^h	St
17 Mon	♈ 11^h	☊7^h	Water/Warmth	Southern Planting Time	Leaf to 5^h, Fruit from 11^h to 15^h	
18 Tue	♈	☉-♑ ♀☋8^h	Warmth	Southern Planting Time	---------------- Fruit from 19^h	St
19 Wed	♉ 9^h		Warmth/Earth	Southern Planting Time	Fruit to 8^h, Root from 9^h	St Vo
20 Thu	♉		Earth	Southern Planting Time	Root	
21 Fri	♉		Earth	Southern Planting Time	Root	
22 Sat	♊ 9^h	⌓20^h	Earth/Light		Root to 8^h, Flower from 9^h	
23 Sun	♊	**Ag** 19^h	Light	NPT	Flower	ϟ
24 Mon	♋ 18^h		Light/Water	NPT	Flower to 17^h, Leaf from 18^h to 24^h	
25 Tue	♋	○11^h ☿☍♄	Water	NPT	Fruit from 1^h to 23^h	ϟ
26 Wed	♌ 12^h		Water/Warmth	NPT	Leaf from 0^h to 11^h, Fruit from 12^h	
27 Thu	♌		Warmth	NPT	Fruit	Tr ϟ
28 Fri	♌	♀☍♄	Warmth	NPT	Fruit	Vo
29 Sat	♍ 9^h		Warmth/Earth	NPT	Fruit to 8^h, Root from 9^h	St
30 Sun	♍		Earth	NPT	Root to 24^h	St
31 Mon	♍	☋23^h	Earth	NPT	---------------- Root from 13^h to 19^h	

Mercury	Venus	Mars	Jupiter	Saturn	Uranus	Neptune	Pluto
♏ 9 ♐ 29 ♑	♏ 8 ♐	♏	♍	♊	♒	♑	♏

***NB:** All zodiac symbols refer to astronomical constellations, not astrological signs (see p.11)*

Notes

1

2
3
4
5
6
7
8

9
10
11
12
13
14
15

16
17
18
19
20
21
22

23
24
25
26
27
28
29

30
31

January 2005

There are very many unfavourable sowing constellations this month. Mercury moves from Scorpion and through Archer to bring some Warmth influences. On Jan 29 it enters the winter constellation of the Goat to bring wintry weather. Venus also starts out from watery Scorpion and on Jan 8 enters the Warmth region of Archer. Mars remains in Scorpion, overtakes Pluto and brings winter rainfall. Jupiter in Virgin and Neptune in Goat are joined by the Sun on Jan 18 to reinforce the Earth element and the winter mood. Saturn in Twins and Uranus in Waterman continue to impart strong Light impulses right up until the summer. Regularly recurring trine positions of Pluto with other planets in Scorpion, will continue to bring bad weather.

Northern Planting Time: Dec 28 to Jan 6 11^h and Jan 22 20^h to Feb 4.

Southern Planting Time: Jan 11 0^h to Jan 21 23^h.

The planting times are also suitable for **pruning** trees, vines and hedges. For fruiting plants select Flower and Fruit days.

Beekeeping notes, see page 44

February 2005

Date	Const. of Moon	Other aspects		Moon Element		Parts of the plant enhanced by Moon or planets	Weather, etc.
1 Tue	♍	☉-♑		Earth	NPT	Root from 0h to 23h	
2 Wed	♎ 0h			Light	NPT	Flower from 0h to 23h	St
3 Thu	♏ 8h			Light/Water	NPT	----------------------------------	
4 Fri	♏			Water	NPT	Leaf from 3h	Vo
5 Sat	♐ 11h	◡22h		Water/Warmth		Leaf to 11h, Fruit from 12h	
6 Sun	♐			Warmth	Southern Planting Time	Fruit to 23h	Eq
7 Mon	♑ 12h	**Pg** 22h	△	Warmth/Earth	Southern Planting Time	Root from 0h to 7h	Tr
8 Tue	♑	● 22h		Earth	Southern Planting Time	Root from 10h	
9 Wed	♒ 8h			Earth/Light	Southern Planting Time	Root to 7h, Flower from 8h to 24h	St Vo
10 Thu	♒		△	Light	Southern Planting Time	Root from 1h to 13h, Flower from 14h to 24h	
11 Fri	♓ 1h			Light/Water	Southern Planting Time	Leaf from 1h	
12 Sat	♓			Water	Southern Planting Time	Leaf to 17h -----------------	Tr
13 Sun	♈ 19h	☊11h		Water/Warmth	Southern Planting Time	Leaf from 0h to 9h, Fruit from 19h	
14 Mon	♈	☉-♒		Warmth	Southern Planting Time	Fruit	
15 Tue	♉ 16h			Warmth/Earth	Southern Planting Time	Fruit to 15h, Root from 16h	St
16 Wed	♉			Earth	Southern Planting Time	Root	
17 Thu	♉			Earth	Southern Planting Time	Root	♄
18 Fri	♊ 16h			Earth/Light	Southern Planting Time	Root to 15h, Flower from 16h	
19 Sat	♊	⌒1h		Light	Northern Planting Time	Flower to 23h	Tr
20 Sun	♊	**Ag** 5h		Light	Northern Planting Time	----------------------------------	St
21 Mon	♋ 0h			Water	Northern Planting Time	Leaf from 9h	
22 Tue	♌ 18h			Water/Warmth	Northern Planting Time	Leaf to 17h, Fruit from 18h	
23 Wed	♌			Warmth	Northern Planting Time	Fruit	St
24 Thu	♌	○5h		Warmth	Northern Planting Time	Fruit	St Tr
25 Fri	♍ 15h			Warmth/Earth	Northern Planting Time	Fruit to 14h, Root from 15h to 23h	
26 Sat	♍			Earth	Northern Planting Time	Root from 10h	
27 Sun	♍		△	Earth	Northern Planting Time	Root to 7h, Flower from 17h to 21h	
28 Mon	♍	☋0h		Earth	Northern Planting Time	Root from 3h	St

Mercury	Venus	Mars	Jupiter	Saturn	Uranus	Neptune	Pluto
♑ 14 ♒ 28 ♓	♑ 23 ♒	♏ 4 ♐	♍	♊	♒	♑	♏

NB: *All zodiac symbols refer to astronomical constellations, not astrological signs* *(see p.11)*

Notes

1
2
3
4
5

6
7
8
9
10
11
12

13
14
15
16
17
18
19

20
21
22
23
24
25
26

27
28

February 2005

Sun and Mercury are in the wintry region of Goat until Feb 14 and this should ensure some frost. This effect will be strengthened by two planetary Earth trines. Venus is in Waterman from Jan 23 and will bring some Light forces. These will be strengthened by a Light trine and an early spring flowering is to be expected.

Northern Planting Time: Jan 22 to Feb 4 23^h and Feb 19 2^h to March 2.

Southern planting time: Feb 5 23^h to Feb 18 23^h.

Vine cuttings and scions for **grafting** should be taken from Feb 13 19^h to Feb 15 15^h and kept cool and moist. If they are to grow as cuttings, they may be inserted in the earth for rooting from Feb 22 18^h to Feb 25 14^h in early districts.

Milk processing

Unfavourable days (– – – –) should be avoided as far as possible for the making of butter and cheese. Milk produced on Warmth-Fruit days will yield the highest butterfat content. This also applies to days with a tendency to thunderstorms. The Moon's perigee (Pg) is nearly always unfavourable for milk processing and even yoghurt will not turn out well. Starters made on these days tend to decay rapidly or curdle easily so that is advisable to make extra the day before. Light and Warmth days offer the best conditions for milk while the Watery Leaf days are unsuitable.

March 2005

Date	Const. of Moon	Other aspects		Moon Element	Parts of the plant enhanced by Moon or planets	Weather, etc.
1 Tue	♎ 5^{h}	☉-♒		Earth/Light	Root to 4^{h}, Flower from 5^{h}	Vo
2 Wed	♏ 14^{h}			Light/Water	Flower to 13^{h}, Leaf from 14^{h} to 19^{h}	St
3 Thu	♏		☿☊11^{h}	Water	----------------- Leaf from 19^{h}	Tr
4 Fri	♐ 19^{h}			Water/Warmth	Leaf to 3^{h} -----------------	
5 Sat	♐	◡5^{h}		Warmth	----------------- Fruit from 18^{h}	Tr Eq
6 Sun	♑ 21^{h}			Warmth/Earth	Fruit to 19^{h}	
7 Mon	♑		♂☍♄	Earth	----------------- Root from 15^{h} to 19^{h}	St
8 Tue	♒ 19^{h}	**Pg** 4^{h}		Earth/Light	----------------- Flower from 19^{h}	
9 Wed	♒			Light	Flower	
10 Thu	♓ 11^{h}	● 9^{h}		Light/Water	Flower	ɥ
11 Fri	♓	☉-♓	△	Water	Flower to 2^{h}, Leaf from 3^{h} to 23^{h}	
12 Sat	♓	☊18^{h}		Water	Leaf from 8^{h} to 16^{h} and from 21^{h}	St
13 Sun	♈ 5^{h}			Water/Warmth	Leaf to 4^{h}, Fruit from 5^{h} to 24^{h}	
14 Mon	♈		△	Warmth	Root from 1^{h} to 9^{h}, Fruit from 10^{h} to 23^{h}	Vo
15 Tue	♉ 1^{h}		△	Warmth/Earth	Flower from 0^{h} to 4^{h}, Root from 5^{h}	
16 Wed	♉			Earth	Root	St
17 Thu	♊ 23^{h}			Earth/Light	Root to 22^{h}, Flower from 23^{h}	
18 Fri	♊	⌒9^{h}		Light	Flower	Vo St
19 Sat	♊	**Ag** 23^{h}		Light	Flower	
20 Sun	♋ 8^{h}			Light/Water	Flower to 7^{h}, Leaf from 8^{h}	St
21 Mon	♋			Water	Leaf	Eq St
22 Tue	♌ 2^{h}			Water/Warmth	Leaf to 1^{h}, Fruit from 2^{h}	Tr
23 Wed	♌			Warmth	Fruit	Eq St
24 Thu	♍ 22^{h}			Warmth/Earth	Fruit to 21^{h}	
25 Fri	♍	*Good Friday*	○21^{h}	Earth	-------------------------------	
26 Sat	♍			Earth	-------------------------------	
27 Sun	♍	*Easter*	☋5^{h}	Earth	Root from 8^{h}	ɥ
28 Mon	♎ 11^{h}			Earth/Light	Root to 10^{h}, Flower from 11^{h}	
29 Tue	♏ 19^{h}			Light/Water	Flower to 24^{h}	ɥ St
30 Wed	♏			Water	Leaf from 1^{h}	
31 Thu	♏			Water	Leaf to 24^{h}	

Southern Planting Time

Northern Planting Time

Mercury	Venus	Mars	Jupiter	Saturn	Uranus	Neptune	Pluto
♓	♒ 15 ♓	♐ 18 ♑	♍	♊	♒	♑	♏

***NB:** All zodiac symbols refer to astronomical constellations, not astrological signs* *(see p.11)*

Notes

1
2
3
4
5

6
7
8
9
10
11
12

13
14
15
16
17
18
19

20
21
22
23
24
25
26

27
28
29
30
31

March 2005

March

In the first half of the month Light and Earth/Water constellations balance one another. From the middle of March Earth and Water constellations dominate. Spring is therefore likely to be delayed. Since the beginning of the year Jupiter is at a nodal point on the Moon's path. This means there will be an occultation when the monthly conjunction occurs as well as a reduction in Jupiter's influence.

Northern Planting Time: Feb 19 to March 2 11^{h} and March 18 11^{h} to March 31 23^{h}.

Southern Planting Time: March 5 18^{h} to March 17 22^{h}. Beware of periods marked as unfavourable (-------).

Moth control period: March 11 12^{h} to March 13 4^{h}.

Cut **scions** March 5 18^{h} to March 6 18^{h}.
Cuttings including those for making willow fences should be planted March 22 2^{h} to March 24 21^{h}.

Sheep awaiting fresh pastures

April 2005

Date	Const. of Moon	Other aspects	Moon Element		Parts of the plant enhanced by Moon or planets	Weather, etc.
1 Fri	♐ 1h	☉-♓ ◡10h	Water/Warmth	Southern Planting Time	Fruit from 1h	
2 Sat	♐	♀☍♃	Warmth		Fruit to 7h ------------------	Vo
3 Sun	♑ 4h	☉☍♃	Warmth/Earth		Fruit from 0h to 19h	St
4 Mon	♑	Pg 11h	Earth		------------------ Root from 19h	♄ St
5 Tue	♒ 3h		Earth/Light		Root to 2h, Flower from 3h	
6 Wed	♓ 21h		Light/Water		Flower to 20h, Leaf from 21h	
7 Thu	♓		Water		Leaf to 18h, Root from 19h	
8 Fri	♓	☉☌ ● 21h △	Water		Root to 6h ----------------	
9 Sat	♈ 15h	☊3h ♀☌	Water/Warmth		---------------------------------	Eq
10 Sun	♈	☿☋18h	Warmth		---------------------------------	♄
11 Mon	♉ 10h	△	Warmth/Earth		Leaf from 0h to 11h, Root from 12h	
12 Tue	♉		Earth		Root	
13 Wed	♉		Earth		Root to 24h	♄
14 Thu	♊ 8h	⌒17h △	Earth/Light	Northern Planting Time	Leaf from 1h to 9h, Flower from 10h	Tr
15 Fri	♊		Light		Flower	Eq Vo Tr
16 Sat	♋ 16h	Ag 19h	Light/Water		Flower to 23h	
17 Sun	♋		Water		Leaf from 0h	Tr
18 Mon	♌ 10h		Water/Warmth		Leaf to 9h, Fruit from 10h	
19 Tue	♌	☉-♈	Warmth		Fruit	Tr
20 Wed	♍ 7h		Warmth/Earth		Fruit to 6h, Root from 7h	
21 Thu	♍		Earth		Root to 23h	
22 Fri	♍	♃☌	Earth		---------------------------------	
23 Sat	♍	☋14h	Earth		Root from 1h to 12h and from 16h to 24h	Vo Tr
24 Sun	♎ 19h	○10h	Earth/Light		---------------------------------	
25 Mon	♎		Light		Flower from 1h	St
26 Tue	♏ 2h		Light/Water		Flower to 1h, Leaf from 2h	
27 Wed	♏		Water		Leaf	St
28 Thu	♐ 7h	◡16h ☿☍♃	Water/Warmth		Leaf to 6h, Fruit from 7h to 20h	
29 Fri	♐	Pg 10h	Warmth		----------------- Fruit from 19h	Eq Vo Tr St
30 Sat	♑ 9h		Warmth/Earth	Southern Planting Time	Fruit to 8h, Root from 9h	St

Mercury	Venus	Mars	Jupiter	Saturn	Uranus	Neptune	Pluto
♓	♓ 15 ♈	♑ 25 ♒	♍	♊	♒	♑	♏

***NB:** All zodiac symbols refer to astronomical constellations, not astrological signs (see p.11)*

Notes

1
2
3
4
5
6
7
8
9
10
11
12
13
14
15
16
17
18
19
20
21
22
23
24
25
26
27
28
29
30

April 2005

Mercury is now retrograde in the watery region of the Fishes. Venus is also in the Fishes in trine position to Pluto in the Scorpion. These Watery constellations dominate the first half of April. Other nodal positions and planetary occultations reinforce this situation. It is to be hoped that from the middle of the month Venus can send some Warmth from the Ram and that during the last week Mars can strengthen the Light forces.

Northern Planting Time: April 14 17^h to April 27 23^h.

Southern Planting Time: April 1 11^h to 14 11^h, and April 29 19^h to May 10. Days with Moon in the Ram are unsuitable for planting potatoes.

Moth control period: April 6 21^h to 9 13^h.

Graft fruit trees and vines: April 24 23^h to 26 1^h, and April 28 17^h to 30 8^h.

Favourable times to pick the preparation plants

Dandelions are picked on Flower days in the morning as soon as they open and while the centre of the flower is still closed. **Yarrow** is picked on Warmth/Fruit days around the time when the Sun stands before the Lion, from the middle of August. **Chamomile** is picked on Light/Flower days just before midsummer. If they are harvested too late the flower heads already begin to form seeds and one is likely to sow chamomile seeds on the fields with the less than successful preparation. They also lose their special properties if there are grubs in the hollow heads which is another reason to harvest at that time. **Stinging nettle** is collected when the first flower buds are opening. The entire plant growing above the earth is harvested on Light/Flower days. **Valerian** is best picked on Light/Flower days during the St. John's period. All the flowers (except valerian) should be laid out on paper and dried in the shade. **Oak bark** is collected on Earth/Root days. The pithy material immediately beneath the bark should not be used.

May 2005

Date	Const. of Moon	Other aspects	Moon Element		Parts of the plant enhanced by Moon or planets	Weather, etc.
1 Sun	♑	☉-♈	Earth	Southern Planting Time	Root	
2 Mon	♒ 8^h		Earth/Light		Root to 7^h, Flower from 8^h	St
3 Tue	♒		Light		Flower	Tr
4 Wed	♓ 3^h		Light/Water		Flower to 2^h, Leaf from 3^h	
5 Thu	♓	*Ascension*	Water		Leaf to 7^h and from 14^h	Tr
6 Fri	♈ 23^h	☊ 11^h	Water/Warmth		Leaf to 9^h and from 14^h to 22^h, Fruit from 23^h	St
7 Sat	♈		Warmth		Fruit to 24^h	Eq
8 Sun	♉ 19^h	● 9^h △	Warmth/Earth		Leaf to 14^h, Fruit 16^h to 18^h, Root from 19^h	St
9 Mon	♉		Earth		Root	♄ Tr
10 Tue	♉		Earth		Root to 19^h ----------------	♄
11 Wed	♊ 16^h	♀☊ 11^h	Earth/Light		--------------------------------	Tr
12 Thu	♊	⌒ 1^h	Light	Northern Planting Time	Flower from 1^h	Tr
13 Fri	♊	☉-♉	Light		Flower	
14 Sat	♋ 0^h	**Ag** 14^h	Water		Flower to 15^h, Leaf from 16^h	
15 Sun	♌ 18^h	*Whitsun*	Water/Warmth		Leaf to 17^h, Fruit from 18^h	
16 Mon	♌		Warmth		Fruit	St
17 Tue	♌	△	Warmth		Fruit to 11^h, Root from 12^h to 24^h	Vo
18 Wed	♍ 16^h		Warmth/Earth		Fruit from 1^h to 15^h, Root from 16^h	♄ Tr
19 Thu	♍	♃•	Earth		Root to 17^h	
20 Fri	♍	☋ 22^h	Earth		Root from 0^h to 20^h	
21 Sat	♍		Earth		Root from 1^h to 23^h	St
22 Sun	♎ 4^h		Earth/Light		----------- Flower from 11^h	Eq Vo
23 Mon	♏ 11^h	○ 20^h	Light/Water		Flower to 10^h, Leaf from 11^h to 23^h	
24 Tue	♏	△	Water		Root from 0^h to 13^h, Leaf from 14^h	Tr St
25 Wed	♐ 14^h		Water/Warmth		Leaf to 13^h, Fruit from 14^h to 21^h	St
26 Thu	♐	**Pg** 11^h ◡ 0^h	Warmth		--------------------------------	
27 Fri	♑ 15^h		Warmth/Earth	SPT	Fruit from 0^h to 14^h, Root from 15^h	
28 Sat	♑		Earth		Root	
29 Sun	♒ 14^h	♀☍♇	Earth/Light		Root to 13^h, Flower from 14^h to 20^h	Tr
30 Mon	♒	△ ☿☊ 10^h	Light		--------------------------------	St
31 Tue	♓ 9^h	♂•	Light/Water		Flower from 0^h to 6^h, Leaf from 13^h	

Mercury	Venus	Mars	Jupiter	Saturn	Uranus	Neptune	Pluto
♓ 11 ♈ 25 ♉	♈ 4 ♉	♒ 30 ♓	♍	♊	♒	♑	♏

NB: *All zodiac symbols refer to astronomical constellations, not astrological signs* *(see p.11)*

Notes

1
2
3
4
5
6
7

8
9
10
11
12
13
14

15
16
17
18
19
20
21

22
23
24
25
26
27
28

29
30
31

May 2005

Mercury is in a Watery region until May 10 and is then in the Ram until May 24. This Warmth impulse is good for all grafting work. Mars and Saturn are in Light constellations. Venus moves into the Bull on May 4 and makes for cooler nights. Potatoes should be planted on Root days. Suitable days for planting potato eyes are May 6 23^h to May 8 18^h. The same applies for planting the tiny potatoes.

Northern Planting Time: May 12 3^h to May 25 13^h.

Southern Planting Time: April 29 to May 10 20^h and from May 26 23^h to June 7.

Straw or wood shavings should now be laid under the **strawberries** to prevent fruit fouling in July.

Grafting in late districts: May 2 8^h to May 4 2^h and May 6 23^h to May 8 18^h.

Control **moths** from May 6 23^h to May 8 18^h.
Control **mole crickets:** May 23 11^h to May 25 13^h.

Take **biodynamic preparations** out of the ground from May 16 0^h to 17 23^h.

Earth warms up from May 11.

Folklore from Germany: If the New Moon in May falls on a Sunday, there'll be a plague of mice. (Solution: Keep more cats.)

June 2005

Date	Const. of Moon	Other aspects	Moon Element		Parts of the plant enhanced by Moon or planets	Weather, etc.
1 Wed	♓	☉-♉	△ Water		Leaf to 5^h, Root 6^h to 13^h, Leaf from16^h	♄ St
2 Thu	♓	☊15^h	Water		Leaf to 12^h and from 17^h	Vo ♄ St
3 Fri	♈ 5^h		Water/Warmth		Leaf to 4^h, Fruit from 5^h	
4 Sat	♈		Warmth	SPT	Fruit to 23^h	
5 Sun	♉ 2^h		△ Water/Earth		Root from 0^h	Eq
6 Mon	♉	● 22^h	Earth		Root	St
7 Tue	♊ 23^h		Earth/Light		Root	
8 Wed	♊	⌒8^h ☿☍♇	△ Light		Root to 7^h, Flower from 8^h	
9 Thu	♊		Light		Flower	St
10 Fri	♋ 7^h		Light/Water		Flower to 6^h, Leaf from 7^h to 23^h	
11 Sat	♋	**Ag** 6^h	Water		Flower from 0^h to 7^h, Leaf from 8^h to 24^h	
12 Sun	♌ 2^h		△ Water/Warmth		Flower from 1^h to 11^h, Fruit from 12^h	Eq ♄ St
13 Mon	♌		Warmth		Fruit	
14 Tue	♌	☉☍♇	Warmth		Fruit to 23^h	♄
15 Wed	♍ 0^h		Earth	Northern Planting Time	Root from 0^h	Eq
16 Thu	♍	♃●	△ Earth		Root to 2^h, Flower 9^h to 19^h, Root from 21^h	St
17 Fri	♍	☋4^h	Earth		Root to 2^h and from 6^h	H
18 Sat	♎ 14^h		Earth/Light		Root to 13^h, Flower from 14^h	
19 Sun	♏ 21^h		Light/Water		Flower to 20^h, Leaf from 21^h	
20 Mon	♏	☉-♊	Water		Leaf	H
21 Tue	♏		Water		Leaf to 23^h	
22 Wed	♐ 0^h	☺4^h ◡8^h	Warmth		Fruit from 0^h to 19^h	
23 Thu	♐	**Pg** 12^h	Warmth		----------- Fruit from 19^h to 23^h	Eq
24 Fri	♑ 0^h		Earth		Root from 0^h	♄
25 Sat	♒ 21^h		Earth/Light		Root to 20^h, Flower from 21^h	♄
26 Sun	♒	♂☍♃	Light	SPT	Flower	Tr
27 Mon	♓ 15^h		Light/Water		Flower to 14^h, Leaf from 15^h	Vo St
28 Tue	♓		Water		Leaf to 17^h	Tr
29 Wed	♓	☊16^h	Water		Leaf from 1^h to 14^h and from 19^h	
30 Thu	♈ 10^h		Water/Warmth		Leaf to 9^h, Fruit from 10^h	

Mercury	Venus	Mars	Jupiter	Saturn	Uranus	Neptune	Pluto
♉ 10 ♊ 26 ♋	♉ 3 ♊ 25 ♋	♓	♍	♊ 22 ♋	♒	♑	♏

***NB:** All zodiac symbols refer to astronomical constellations, not astrological signs (see p.11)*

Notes

1
2
3
4

5
6
7
8
9
10
11

12
13
14
15
16
17
18

19
20
21
22
23
24
25

26
27
28
29
30

June 2005

The month will start cool and moist. Mars is in the Watery region of Fishes throughout June. Venus enters the Light region of the Twins on June 3. This will give good quality hay and is favourable to increasing the number of bee colonies. On June 22 Saturn enters the Watery region of the Crab. Venus and Mercury join him on June 25 and 26. There will then be five planets in Water constellations — a delight for slugs and snails! It will be worthwhile on June 26 to stir Horn Silica for an hour early in the morning and spray the entire garden including lawns and pathways with it.

Northern Planting Time: June 8 10^{h} to June 21 21^{h}.

Southern Planting Time: May 26 to June 7 23^{h} and June 24 1^{h} to July 4.

Insect pest control: June 5 2^{h} to June 7 19^{h}.
To control **flies** in the cow shed burn fly papers on Flower days between June 3 and 24.

Good days for **haymaking:** June 3 to 24 preferably on Flower days. Specially good for cutting on June 12 0^{h} to 24^{h}.

Dealing with fungus attacks on plants

The role of fungi in nature is to break down and decompose life that is dying. They appear on our cultivated plants if unripe manure compost or uncomposted animal by-products such as hoof and horn meal are used, but also if the seeds are harvested during unfavourable constellations, or according to Rudolf Steiner: "when the moon forces are working too strongly on the earth." A tea can be made from the leaves and stems of horsetail (*equisetum arvense*) which is then sprayed on the soil around affected plants. By doing so the fungus niveau will be drawn down into the soil where it really belongs.

The health of the plants can be improved by spraying stinging nettle tea on to the leaves. This will stimulate the plants metabolism, encourage the flow of sap and consequently cause the fungus to disappear.

July 2005

Date	Const. of Moon		Other aspects		Moon Element		Parts of the plant enhanced by Moon or planets	Weather, etc.
1 Fri	♈		☉-♊		Warmth	SPT	Fruit	♄ St
2 Sat	♉	7h		△	Warmth/Earth	SPT	Fruit to 6h, Flower from 7h to 18h, Root from 19h	St
3 Sun	♉				Earth	SPT	Root	Vo
4 Mon	♉				Earth	SPT	Root	Vo
5 Tue	♊	6h	☊ 13h		Earth/Light	Northern Planting Time	Root to 5h, Flower from 6h	St
6 Wed	♊		● 12h		Light	Northern Planting Time	Flower to 24h	♄ St
7 Thu	♋	14h		☿☋17h	Light/Water	Northern Planting Time	--------------------------	UW
8 Fri	♋		**Ag** 18h		Water	Northern Planting Time	Leaf from 0h to 14h, Flower from 15h to 23h	H ♄ Tr
9 Sat	♌	8h			Water/Warmth	Northern Planting Time	Leaf from 0h to 7h, Fruit from 8h	
10 Sun	♌				Warmth	Northern Planting Time	Fruit	
11 Mon	♌				Warmth	Northern Planting Time	Fruit	
12 Tue	♍	7h		♀☍♆	Warmth/Earth	Northern Planting Time	Fruit to 6h, Root from 7h	
13 Wed	♍		♃●	☿☍♆	Earth	Northern Planting Time	Root to 13h and from 20h	
14 Thu	♍		☋7h		Earth	Northern Planting Time	Root to 5h and from 10h to 23h	
15 Fri	♎	23h		△	Earth/Light	Northern Planting Time	Leaf 0h to 19h, Root 20h to 22h, Flower from 23h	
16 Sat	♎				Light	Northern Planting Time	Flower	
17 Sun	♏	7h			Light/Water	Northern Planting Time	Flower to 6h, Leaf from 7h	♄
18 Mon	♏				Water	Northern Planting Time	Leaf	
19 Tue	♐	11h	☉-♋	◡18h	Water/Warmth		Leaf to 10h, Fruit from 11h	
20 Wed	♐				Warmth	Southern Planting Time	Fruit to 24h	♄ St
21 Thu	♑	10h	☺11h	**Pg** 20h	Warmth/Earth	Southern Planting Time	--------------------------	
22 Fri	♑				Earth	Southern Planting Time	Root from 8h	Tr
23 Sat	♒	6h			Earth/Light	Southern Planting Time	Root to 5h, Flower from 6h	
24 Sun	♓	23h			Light/Water	Southern Planting Time	Flower to 22h, Leaf from 23h	
25 Mon	♓				Water	Southern Planting Time	Leaf	Tr
26 Tue	♓		☊18h		Water	Southern Planting Time	Leaf to 16h and from 21h	♄ St
27 Wed	♈	17h			Water/Warmth	Southern Planting Time	Leaf to 16h, Fruit from 17h	
28 Thu	♈				Warmth	Southern Planting Time	Fruit	St
29 Fri	♉	13h			Warmth/Earth	Southern Planting Time	Fruit to 12h, Root from 13h	St
30 Sat	♉				Earth	Southern Planting Time	Root to 24h	Vo
31 Sun	♉			♀☍⛢	Earth	Southern Planting Time	--------------------------	

Mercury	Venus	Mars	Jupiter	Saturn	Uranus	Neptune	Pluto
♋ 15 ♌	♋	♓	♍	♋	♒	♑	♏
31 ♋	13 ♌	26 ♈					

***NB:** All zodiac symbols refer to astronomical constellations, not astrological signs (see p.11)*

Notes

1
2

3
4
5
6
7
8
9

10
11
12
13
14
15
16

17
18
19
20
21
22
23

24
25
26
27
28
29
30

31

July 2005

During the first half of the month the Watery element will be dominant. After that Venus and Mercury enter the Warmth region of the Lion and Mars moves into the Ram. This should bring some hope of warmer weeks. Flower and Fruit days are suitable for the summer pruning of fruit trees. Berries for making jams or for drying should be harvested on Flower or Fruit days.

Northern Planting Time: July 5 14^h to July 18 23^h.
Southern Planting Time: June 21 to July 4 23^h and July 19 19^h to Aug 1.

Control **grasshoppers** and **locusts:** July 5 7^h to July 7 13^h.
Control **snails** and **slugs:** July 19 to August 9 on all Water days and especially when the Moon is in the Crab (Aug 3–4).

Harvest cereals: Second half of July on Fruit days.

Supplementary feeding of bees

Herb teas are recommended as a supplement to winter feeding. They have proved their worth over many years as regards the health of colonies. Yarrow, chamomile, dandelion and valerian are used by making an infusion from the dried flowers with boiling water and straining it after 15 minutes. Nettles, equisetum and oak bark are put into cold water, brought to the boil and left to stand for ten minutes before straining. These teas are added to the liquid feed. Three grams of each herb tea is sufficient for 100 litres of liquid feed. This is particularly important in years when there is much honey dew.

August 2005

Date	*Const. of Moon*	*Other aspects*		*Moon Element*		*Parts of the plant enhanced by Moon or planets*	*Weather, etc.*
1 Mon	♊ 11h	☉-♋	⌒18h	Earth/Light		---------------- Flower from 11h	
2 Tue	♊		☿☍♆	Light	Northern Planting Time	Flower	
3 Wed	♋ 20h			Light/Water	Northern Planting Time	Flower to 19h, Leaf from 20h	
4 Thu	♋	**Ag** 22h		Water	Northern Planting Time	Leaf to 17h, Flower from 18h to 24h	
5 Fri	♌ 14h	● 3h		Water/Warmth	Northern Planting Time	Leaf from 0h to 13h, Fruit from 14h	**H**
6 Sat	♌			Warmth	Northern Planting Time	Fruit	
7 Sun	♌			Warmth	Northern Planting Time	Fruit to 23h	**St**
8 Mon	♍ 13h	☉☍♆	♀☌	Warmth/Earth	Northern Planting Time	---------------- Root from 13h	
9 Tue	♍			Earth	Northern Planting Time	Root to 24h	**Vo**
10 Wed	♍	☉-♌	☋8h ♃☌	Earth	Northern Planting Time	---------------- Root from 11h	**Vo Eq St**
11 Thu	♍			Earth	Northern Planting Time	Root	
12 Fri	♎ 6h			Earth/Light	Northern Planting Time	Root to 5h, Flower from 6h	**♄ St**
13 Sat	♏ 15h			Light/Water	Northern Planting Time	Flower to 14h, Leaf from 15h	
14 Sun	♏			Water	Northern Planting Time	Leaf	**♄ St**
15 Mon	♐ 20h			Water/Warmth	Northern Planting Time	Leaf to 19h, Fruit from 20h	
16 Tue	♐	◡3h		Warmth	Southern Planting Time	Fruit	
17 Wed	♑ 21h		△	Warmth/Earth	Southern Planting Time	Fruit to 11h, Root from 12h	**Eq**
18 Thu	♑			Earth	Southern Planting Time	Root to 17h	**St**
19 Fri	♒ 17h	○18h	**Pg** 6h	Earth/Light	Southern Planting Time	---------------- Flower from 17h	
20 Sat	♒			Light	Southern Planting Time	Flower	
21 Sun	♓ 9h			Light/Water	Southern Planting Time	Flower to 8h, Leaf from 9h	**Tr**
22 Mon	♓	☊23h		Water	Southern Planting Time	Leaf to 7h and from 12h to 21h	
23 Tue	♓			Water	Southern Planting Time	Leaf from 2h to 24h	**St**
24 Wed	♈ 1h			Water/Warmth	Southern Planting Time	Fruit from 1h	
25 Thu	♉ 20h			Warmth/Earth	Southern Planting Time	Fruit to 19h	**Tr**
26 Fri	♉	☿☊9h	☿☍♆	Earth	Southern Planting Time	--------------------------------	
27 Sat	♉			Earth	Southern Planting Time	Root from 1h	
28 Sun	♊ 17h			Earth/Light	Southern Planting Time	Root to 16h, Flower from 17h	**Eq St**
29 Mon	♊	⌒0h		Light	Northern Planting Time	Flower to 23h	**♄**
30 Tue	♊		△	Light	Northern Planting Time	Root from 0h to 11h ------------	
31 Mon	♋ 2h		♀☋0h	Light/Water	Northern Planting Time	--------------------------------	

Mercury	Venus	Mars	Jupiter	Saturn	Uranus	Neptune	Pluto
♋ 28 ♌	♌ 11 ♍	♈	♍	♋	♒	♑	♏

***NB:** All zodiac symbols refer to astronomical constellations, not astrological signs (see p.11)*

Notes

1
2
3
4
5
6

7
8
9
10
11
12
13

14
15
16
17
18
19
20

21
22
23
24
25
26
27

28
29
30
31

August 2005

With Mercury and Saturn in the Crab and Pluto still in the Scorpion, the Watery element is well represented. During the first ten days Mars in the Ram and Venus in the Lion hopefully provides some counterweight. On Aug 10 Venus moves away from the Sun towards the Virgin to encourage the first autumn mists. August will not bring the extremes we have seen in recent years. The normally favourable effects of planetary oppositions will be somewhat compromised by their nodal positions and occultations.

Northern Planting Time: Aug 1 20^h to Aug 15 19^h and Aug 29 2^h to Sep 11.

Southern Planting Time: July 19 to Aug 1 17^h and Aug 16 11^h to Aug 28 23^h.

Early **apple** crops should be harvested on Flower and Fruit days.

Favourable times for **harvesting cereals:** Aug 28 to Sep 15 on Fruit days.

Harvesting and processing

Unfavourable times (– – – –) as well as Leaf days are best avoided when harvesting fruits which are to be stored or conserved in any way. The same is true for their subsequent processing whether into juice, jam, jelly or for baking. It also applies when making lactic-acid-fermented vegetables and sauerkraut.

Juice from **plums**, **cherries**, **rose hips, apples** and **pears** can be boiled down to produce a wonderful jams. If, however, these fruits are harvested at the above mentioned times, the spread will rapidly go mouldy. These times should also be avoided when picking the leaves and flowers of herbs for teas otherwise they will have little flavour. Flower and Fruit days should be chosen in all the above cases.

If unfavourable times (– – – –) are used to harvest **cereals** then the following year ergot in rye and bunt in wheat may be expected.

September 2005

Date	Const. of Moon	Other aspects	Moon Element		Parts of the plant enhanced by Moon or planets	Weather, etc.
1 Thu	♌ 20^h	☉-♌ **Ag** 3^h	Water/Warmth	Northern Planting Time	Fruit and Flower! from 0^h to 23^h	☉☍♅ **Tr**
2 Fri	♌		Warmth		Fruit	
3 Sat	♌	● 19^h	Warmth		Fruit	
4 Sun	♍ 18^h		Warmth/Earth		Fruit to 17^h, Root from 18^h	**St**
5 Mon	♍		Earth		Root to 14^h ----------------	♄ **St**
6 Tue	♍	☋ 11^h	Earth		Root from 1^h to 9^h and from 14^h	
7 Wed	♍	♀●	Earth		Root to 3^h and from 11^h	**H St**
8 Thu	♎ 12^h		Earth/Light		Root to 11^h, Flower from 12^h to 23^h	
9 Fri	♏ 21^h	☿☍♅	Light/Water		--------------------------------	**Vo**
10 Sat	♏		Water		---------------- Leaf from 12^h	**Tr**
11 Sun	♏		Water		Leaf	**St**
12 Mon	♐ 4^h	◡ 11^h	Water/Warmth	Southern Planting Time	Leaf to 3^h ----------------	**UW**
13 Tue	♐	△	Warmth		Fruit from 0^h	**St**
14 Wed	♑ 6^h		Warmth/Earth		Fruit to 5^h, Root from 6^h	**Vo**
15 Thu	♑	☉-♍	Earth		Root to 24^h	**St**
16 Fri	♒ 3^h	**Pg** 14^h	Earth/Light		--------------------------------	**Vo**
17 Sat	♓ 20^h		Light/Water		Flower from 0^h to 19^h, Leaf from 20^h	**Vo**
18 Sun	♓	☺ 2^h △	Water		Leaf to 4^h, Flower from 5^h to 19^h, Leaf from 20^h	
19 Mon	♓	☊ 9^h	Water		Leaf to 6^h and from 10^h to 13^h	
20 Tue	♈ 11^h		Water/Warmth		Leaf from 0^h to 10^h, Fruit from 11^h	**Eq**
21 Wed	♈		Warmth		Fruit	
22 Thu	♉ 5^h		Warmth/Earth		Fruit to 4^h, Root from 5^h	
23 Fri	♉		Earth		Root	**Eq** ♄
24 Sat	♉		Earth		Root to 24^h	**Eq**
25 Sun	♊ 1^h	◠ 7^h	Earth/Light	NPT	Flower from 1^h	**Vo St**
26 Mon	♊		Light		Flower	**Tr**
27 Tue	♋ 8^h		Light/Water		Flower to 7^h, Leaf from 8^h	**St**
28 Wed	♋	**Ag** 15^h	Water		Leaf to 11^h, Flower 12^h to 17^h, Leaf 18^h to 24^h	♄
29 Thu	♌ 3^h	△	Water/Warmth		Root from 1^h to 10^h, Fruit from 11^h	
30 Fri	♌		Warmth		Fruit	**St**

Mercury	Venus	Mars	Jupiter	Saturn	Uranus	Neptune	Pluto
♌	♍	♈	♍	♋	♒	♑	♏
16 ♍	19 ♎	24 ♉					

***NB:** All zodiac symbols refer to astronomical constellations, not astrological signs (see p.11)*

Notes

1
2
3

4
5
6
7
8
9
10

11
12
13
14
15
16
17

18
19
20
21
22
23
24

25
26
27
28
29
30

September 2005

For the first two weeks summer weather can be expected and then autumn will arrive. Venus is in the Light constellation of the Scales from Sep 19 and will lift some of the mist during the day. In the middle of the month Warmth and Light trines will bring summer weather for planting cereals. In the northern hemisphere the Moon is ascending at this time and it is a good time for harvesting early fruit varieties for storage. In the second half of the month Earth constellations predominate and these will bring cool nights but good conditions for harvesting roots.

Northern Planting Time: Aug 29 to Sep 11 23^h and Sep 25 11^h to Oct 9.

Southern Planting Time: Sep 12 14^h to Sep 24 23^h.

Control **ants in buildings:** Sep 1 20^h to Sep 4 17^h.

For **fruit harvesting** the best days are when the Moon is in Aries or Sagittarius. Recent storage tests have confirmed this. Again new storage tests confirmed that the harvest of roots such as onions, carrots, beetroot and potatoes on Root days give much better results.

October 2005

Date	Const. of Moon	Other aspects	Moon Element		Parts of the plant enhanced by Moon or planets	Weather, etc.
1 Sat	♌	☉-♍	Warmth	Northern Planting Time	Fruit	
2 Sun	♍ 1^h	♀☍♂	Warmth/Earth		Fruit to 11^h, Root from 12^h to 23^h	Tr
3 Mon	♍	● 10^h ☋17^h ☉☌	Earth		----------------------------------	☿☋16^h ♄ Tr
4 Tue	♍	☿☌	Earth		----------------------------------	
5 Wed	♎ 17^h		Earth/Light		Root from 0^h to 16^h, Flower from 17^h	
6 Thu	♎		Light		Flower	♄ St
7 Fri	♏ 2^h		Light/Water		Flower to 1^h, Leaf from 2^h to 17^h, Root from 18^h	
8 Sat	♏	△	Water		Root to 7^h, Leaf from 8^h	
9 Sun	♐ 10^h	◡17^h	Water/Warmth		Leaf to 9^h, Fruit from 10^h	
10 Mon	♐		Warmth	Southern Planting Time	Fruit	Tr St
11 Tue	♑ 13^h		Warmth/Earth		Fruit to 12^h, Root from 13^h	
12 Wed	♑		Earth		Root	
13 Thu	♒ 12^h		Earth/Light		Root to 11^h, Flower from 12^h to 23^h	Vo
14 Fri	♒	**Pg** 14^h	Light		----------------------------------	♄
15 Sat	♓ 5^h		Light/Water		Flower from 0^h to 4^h, Leaf from 5^h	♄ Tr
16 Sun	♓	☊19^h	Water		Leaf to 16^h and from 21^h	
17 Mon	♈ 21^h	○☍12^h △	Water/Warmth		Leaf to 7^h and from 14^h to 20^h, Fruit from 21^h	
18 Tue	♈		Warmth		Fruit to 16^h ------------------	Eq
19 Wed	♉ 15^h		Warmth/Earth		Fruit from 2^h to 14^h, Root from 15^h	St
20 Thu	♉		Earth		Root	
21 Fri	♉		Earth		Root to 23^h	
22 Sat	♊ 9^h	◠15^h ☿☍♂	Earth/Light		Fruit from 0^h to 24^h	
23 Sun	♊		Light	Northern Planting Time	Flower from 1^h	
24 Mon	♋ 16^h		Light/Water		Flower to 15^h, Leaf from 16^h	
25 Tue	♋		Water		Leaf to 23^h	
26 Wed	♌ 11^h	**Ag** 10^h	Water/Warmth		Fruit from 0^h	Tr
27 Thu	♌		Warmth		Fruit	
28 Fri	♌		Warmth		Fruit	
29 Sat	♍ 9^h		Warmth/Earth		Fruit to 8^h, Root from 9^h	
30 Sun	♍		Earth		Root to 23^h	Vo St
31 Mon	♍	☋1^h	Earth		Root from 3^h	Tr

Mercury	Venus	Mars	Jupiter	Saturn	Uranus	Neptune	Pluto
♍ 14 ♎ 27 ♏	♎ 5 ♏	♉ 9 ♈	♍	♋	♒	♑	♏

***NB:** All zodiac symbols refer to astronomical constellations, not astrological signs (see p.11)*

Notes

1

2
3
4
5
6
7
8

9
10
11
12
13
14
15

16
17
18
19
20
21
22

23
24
25
26
27
28
29

30
31

October 2005

Mars is retrograde in the Ram where he has his greatest influence. He is in opposition to Mercury and Venus at the end of the month. This will provide a good balance to the Earth and Water constellations of the other planets.

Northern Planting Time: Sep 25 to Oct 9 8^h and Oct 22 17^h to Nov 5.

Southern Planting Time: Oct 9 23^h to Oct 22 11^h.

Crops for storage: Harvest fruit on Flower and Fruit days except during planting time. Roots are best harvested on Root days preferably during planting time. Cereals should be planted on Fruit days.

Harvested areas can be given compost and left ploughed over winter.

Burning the skins of **red-blooded animal pests** should be carried out on Oct 20.

Making rye bread

Wheat, **barley**, **oats**, **maize**, **rice** and **millet** can be milled into flour to make very good bread using yeast or a rising agent. With **rye** however another procedure is necessary. A detailed recipe for making rye bread was published in last year's calendar. Rye needs a room temperature of over 20°C. I was amazed to see how during the heatwave of 2003 the whole sequence of five risings was completed in the space of four hours. This means of course that the rising of sour dough is largely dependent on warmth. Incidentally, Elsewhere Finnish scientists have shown that the eating of rye bread has helped cancer patients recover.

Oct

November 2005

Date	Const. of Moon	Other aspects		Moon Element		Parts of the plant enhanced by Moon or planets	Weather, etc.
1 Tue	♍	☉-♎		Earth	NPT	Root to 24h	Tr
2 Wed	♎ 1h	● 1h		Earth/Light	NPT	Flower from 1h	Vo St
3 Thu	♏ 9h			Light/Water	NPT	Flower to 8h, Leaf from 9h	
4 Fri	♏			Water	NPT	Leaf	
5 Sat	♐ 15h	◡ 22h		Water/Warmth	NPT	Leaf to 14h, Fruit from 15h to 23h	Eq
6 Sun	♐			Warmth		----------------------------------	Eq ♄
7 Mon	♑ 19h		☉☍♂	Warmth/Earth	Southern Planting Time	Fruit from 10h to 18h, Root from 19h	UW Eq
8 Tue	♑			Earth	Southern Planting Time	Root	St
9 Wed	♒ 18h			Earth/Light	Southern Planting Time	Root to 11h -----------------	
10 Thu	♒	**Pg** 0h		Light	Southern Planting Time	----------------- Flower from 14h	St
11 Fri	♓ 13h			Light/Water	Southern Planting Time	Flower to 12h, Leaf from 13h	
12 Sat	♓			Water	Southern Planting Time	Leaf to 23h	
13 Sun	♓	☊ 2h		Water	Southern Planting Time	Leaf from 4h	
14 Mon	♈ 6h			Water/Warmth	Southern Planting Time	Leaf to 5h, Fruit from 6h to 19h	St
15 Tue	♈		♂☊ 11h	Warmth	Southern Planting Time	----------------------------------	
16 Wed	♉ 0h	○ 1h		Earth	Southern Planting Time	Root from 0h to 24h	St
17 Thu	♉		△	Earth	Southern Planting Time	Fruit from 1h to 23h	
18 Fri	♊ 18h			Earth/Light	Southern Planting Time	Root from 0h to 17h, Flower from 18h	
19 Sat	♊	☉-♏	⌒ 0h	Light	Northern Planting Time	Flower	St
20 Sun	♊			Light	Northern Planting Time	Flower to 24h	
21 Mon	♋ 1h			Light/Water	Northern Planting Time	Leaf from 1h to 17h	♄
22 Tue	♌ 19h		☿☊ 9h	Water/Warmth	Northern Planting Time	----------------------------------	
23 Wed	♌	**Ag** 6h		Warmth	Northern Planting Time	Fruit from 0h to 14h -----------------	
24 Thu	♌			Warmth	Northern Planting Time	----------------- Fruit from 17h	Tr St
25 Fri	♍ 18h			Warmth/Earth	Northern Planting Time	Fruit to 17h, Root from 18h	Tr
26 Sat	♍			Earth	Northern Planting Time	Root	
27 Sun	♍	☋ 9h		Earth	Northern Planting Time	Root to 5h and from 10h	
28 Mon	♍			Earth	Northern Planting Time	Root	♄
29 Tue	♎ 10h			Earth/Light	Northern Planting Time	Root to 9h, Flower from 10h	♄
30 Wed	♏ 18h			Light/Water	Northern Planting Time	Flower to 17h, Leaf from 18h	♄

Mercury	Venus	Mars	Jupiter	Saturn	Uranus	Neptune	Pluto
♏ 27 ♎	♏ 3 ♐	♈	♍	♋	♒	♑	♏

***NB:** All zodiac symbols refer to astronomical constellations, not astrological signs (see p.11)*

Notes

1
2
3
4
5

6
7
8
9
10
11
12

13
14
15
16
17
18
19

20
21
22
23
24
25
26

27
28
29
30

November 2005

Mercury is retrograde in the Scorpion from Nov 14 and is near to Pluto and in trine position with Saturn in the Crab. Heavy precipitation is to be expected. Venus in the Archer and Mars in the Ram will provide a suitable window around Nov 17 for completing the fruit and wine harvest and the delayed cereal plantings.

Northern Planting Time: Oct 22 17^h to Nov 5 11^h and Nov 18 2^h to Dec 2.

Southern Planting Time: Nov 7 11^h to Nov 18 21^h.

Flower and Fruit days are good for **pruning** fruit and vines as well as for felling **timber.** Special woods like oak, yew, ash, spruce and fir should be felled on Nov 8.
Flower bulbs: Plant on Flower days until Nov 20.
Christmas trees should be cut for sale on Flower and Fruit days during the ascending Moon from Nov 7 to 18.

Tipped manure generally rots better than manure which has been through a spreader.

December 2005

Date	Const. of Moon	Other aspects	Moon Element	Planting time	Parts of the plant enhanced by Moon or planets	Weather, etc.
1 Thu	♏	☉-♏ ●15^h	Water		Leaf	Vo
2 Fri	♐ 23^h		Water/Warmth		Leaf	
3 Sat	♐	⌣5^h △	Warmth	Southern Planting Time	Leaf to 8^h, Fruit from 9^h	
4 Sun	♐		Warmth	Southern Planting Time	Fruit to 17^h	
5 Mon	♑ 1^h	**Pg** 5^h ♂☍♃	Warmth/Earth	Southern Planting Time	---------------- Root from 17^h	
6 Tue	♒ 23^h		Earth/Light	Southern Planting Time	Root to 22^h, Flower from 23^h	St
7 Wed	♒		Light	Southern Planting Time	Flower	Vo
8 Thu	♓ 18^h		Light/Water	Southern Planting Time	Flower to 17^h, Leaf from 18^h	
9 Fri	♓		Water	Southern Planting Time	Leaf	
10 Sat	♓	☊5^h	Water	Southern Planting Time	Leaf to 3^h and from 8^h	♄
11 Sun	♈ 12^h		Water/Warmth	Southern Planting Time	Leaf to 11^h, Fruit from 12^h to 19^h	
12 Mon	♈	♂☌	Warmth	Southern Planting Time	Fruit from 7^h	
13 Tue	♉ 8^h		Warmth/Earth	Southern Planting Time	Fruit to 7^h, Root from 8^h	
14 Wed	♉		Earth	Southern Planting Time	Root	
15 Thu	♉	○16^h	Earth	Southern Planting Time	Root	**Tr St**
16 Fri	♊ 3^h	⌢7^h	Earth/Light	Northern Planting Time	Root to 2^h, Flower from 3^h	**St**
17 Sat	♊		Light	Northern Planting Time	Flower	**UW Tr**
18 Sun	♋ 9^h		Light/Water	Northern Planting Time	Flower to 8^h, Leaf from 9^h	
19 Mon	♋	☉-♐	Water	Northern Planting Time	Leaf	♄
20 Tue	♌ 3^h		Water/Warmth	Northern Planting Time	Leaf to 2^h, Fruit from 3^h to 23^h	
21 Wed	♌	**Ag** 3^h △	Warmth	Northern Planting Time	Leaf from 0^h to 13^h ----------------	
22 Thu	♌	♀☊4^h	Warmth	Northern Planting Time	---------------- Fruit from 17^h	**Eq St**
23 Fri	♍ 2^h		Warmth/Earth	Northern Planting Time	Fruit to 1^h, Root from 2^h	
24 Sat	♍	☋11^h	Earth	Northern Planting Time	Root to 9^h and from 14^h	**St**
25 Sun	♍		Earth	Northern Planting Time	Root	**Vo St**
26 Mon	♎ 19^h		Earth/Light	Northern Planting Time	Root to 13^h ----------------	
27 Tue	♎		Light	Northern Planting Time	Flower from 0^h	**St**
28 Wed	♏ 4^h		Light/Water	Northern Planting Time	Flower to 3^h, Leaf from 4^h	
29 Thu	♏		Water	Northern Planting Time	Leaf to 23^h	**St**
30 Fri	♐ 9^h	⌣14^h ☿☋16^h	Water/Warmth		------------------------------	
31 Sat	♐	●3^h	Warmth		Fruit from 0^h	**Eq St**

Mercury	Venus	Mars	Jupiter	Saturn	Uranus	Neptune	Pluto
♎	♐	♈	♍	♋	♒	♑	♏
9 ♏	10 ♏		8 ♎				

***NB:** All zodiac symbols refer to astronomical constellations, not astrological signs* *(see p.11)*

Notes

1
2
3

4
5
6
7
8
9
10

11
12
13
14
15
16
17

18
19
20
21
22
23
24

25
26
27
28
29
30
31

December 2005

Mercury and Venus are in Light and Warmth constellations during the first ten days. Mars continues to be retrograde in the Ram. From Dec 10 Venus moves into the wintery constellation of the Goat bringing clear night skies. On Dec 8 Jupiter moves into the Light constellation of the Scales, and begins to manifest his Light impulse and in the new years come into trine position with Uranus in Waterman.

Northern Planting Time: Nov 18 2^h to Dec 2 20^h and from Dec 16 9^h to Dec 29 23^h.

Southern Planting Time: Dec 3 11^h to Dec 15 23^h and from Dec 31 0^h.

This is a good time to **prune** trees, vines and hedges. Choose Flower and Fruit days. Fell special woods like oak, yew, apple and maple on Dec 6.

Christmas trees: Cut during ascending Moon on Fruit and Blossom days: Dec 2 23^h to Dec 5 2^h, and from Dec 6 23^h to Dec 8 16^h, and from Dec 11 12^h to Dec 13 8^h.

We would like to wish all our readers
a peaceful Christmas
with lots of health and strength for the New Year

Animal and Insect pests

The first question to ask is: Why does a particular animal appear as a pest? The first step is to familiarize oneself with the living conditions and habits of the creature in question and to rectify any management errors that have been made. If despite this an animal becomes a pest, it can be driven back to within its natural limits using its own burnt remains — its ashes. For detailed recommendations see the monthly notes.

There is no need to reach for biological and chemical means of control; instead regulate the animal or insect from within its own species. For mice, birds and the like, a few skins or the feathers should suffice while for insects, slugs etc. the following method should be followed:

Take 50 or 60 specimens of the particular pest and burn them in a wood fire during the appropriate planetary aspect. The resulting mixture of wood and pest ash should then be "dynamized" by grinding it in a mortar for an hour.

One gram of this dynamized ash should then be placed in a small bottle with 9 cc (grams) of water and shaken vigorously for three minutes. This is the first decimal potency (D1 or X1). A further 90 cc of water are then added and it is again shaken for three minutes. This is the second decimal potency, D2 or X2. Repeating this procedure until D8 (X8) would produce 100,000 litres! It is therefore advisable to proceed until D4 and then start again using smaller quantities (1 cc of D4 in 9 cc water).

This D8 potency has been found to exhibit an inhibiting effect on the reproductive capacity of the pest when it is applied as a fine mist for three evenings in succession. Good results have been reported for several species. Where pests occur in large numbers good results are obtained by burning them on the site where they have been found. Flea beetle and apple blossom weevil can be caught with fly papers for example and burnt on site.

Sowing times for trees and shrubs

Jan 27	Plum, Juniper, Thuja, Blackthorn, Birch, Pear, Lime, Sweet cherry
March 7	Hornbeam, Rowan, Gean, Oak, Yew, Quince
April 2	Birch, Pear, Lime, Apple, Maple , Beech
April 3	Crab apple, Maple, Ash, Hawthorn, Spruce, Walnut, Horse chestnut
April 28	Elm, Grape vine, Crab apple, Peach, Horse chestnut, Beech
June 26	Oak, Yew, Rowan, Sour cherry, Quince, Maple, Apple
July 31	Pear, Birch, Lime, Mirabelle plum, Raspberry
Sep 1	Hazel, Spruce, Fir, Cedar, Olive, Walnut, Ash
Sep 9	Larch, Elm, Grape vine, Elder
Oct 2	Birch, Lime, Pear, Oak, Yew, Sour cherry
Oct 22	Elder, Larch, Elm, Rowan
Nov 7	Hazel, Sweet chestnut, Horse chestnut, Oak, Yew, Ash, Walnut
Dec 5	Oak, Maple, Apple, Sour cherry, Wild rose

Left: Maple seeds *Top: Walnuts* *Bottom: Two generations of fir cones*

Beekeeping

Bees live in skeps or hives shut away from the outside world. As extra protection they line their dwelling with propolis to keep harm away. Direct connection with the surrounding world is maintained by the flying bees.

If the beekeeper wants the colony to take advantage of cosmic forces he must do something similar to what the nurseryman does when cultivating the soil. With the air, cosmic forces penetrate the soil which are then taken in by the plant and can be made use of until the next cultivation. The beekeeper must open up the beehive or break the propolic seal. This causes disturbance through which the cosmic forces find the way into the bee colony and can be effective until the next operation. In this way the beekeeper can quite deliberately direct cosmic influences to the bees.

Now it is not a matter of indifference which influences from the surroundings are invoked by the precise time when the bees are handled. Here the beekeeper can quite consciously intervene and use the appropriate days to encourage particular activities such as building up the colony, gathering food according to the needs of the colony at any particular stage of its development. The colony rewards the beekeeper and gives him back part of the honey made from the materials gathered and transformed. Thus Earth-Root days are recommended when the colony needs to build comb. Light-Flower days stimulate brood-rearing and the building up of the colony. Handling on Warmth-Fruit days stimulates the desire to make honey. Water-Leaf days are unsuitable for handling or taking and extracting honey.

Since the end of the seventies the varroa mite has spread into most European countries. After many trials and comparisons of various ways of using the ashes, we recommend burning the mites as described on page 42, to dynamize the ashes by pulverizing them in a mortar for an hour. Then finely strew them with a salt sifter in the spaces between the frames. The preparation and use of these ashes should be done when the Sun and Moon are in the Bull.

Monthly notes for beekeeping

January: Because of damp conditions during the first weeks of January the bees will remain quietly dormant. Later planetary constellations will be less peaceful.

February: On Flower and Fruit days the bees will make cleansing flights although nights will still be cold during the first half of the month.

March: While the first half of March brings agreeable conditions for the bees, the second half could cause some setbacks especially in colder areas. Food reserves need to be checked carefully.

April: Only after April 15 can better bee weather be expected. The colonies should still be kept warm.

May: From May 11 onward the long-awaited spring should arrive for the bees.

June: Make use of Flower days from June 3 for increasing colonies and breeding queens. Such breeding work should be completed by June 22.

July: In the second half of the month there are some good honey constellations. If an adequate amount of intermediate crops have been planted, the bees should be able to find a plentiful supply to stock up for winter.

August: Feeding for winter should be completed this month. Further smaller honey flows are possible from August 28. The autumn promises to make up for the poor spring.

September: The position of Mercury and Mars will encourage a flow of nectar in the first half of the month and from Sep 19 Venus brings a further pollen crop.

October: Further small nectar flows are possible in October if the bees find sufficient flowers or if the right catch and fodder crops have been planted.

November: November is very strongly influenced by Venus and Mars and there will be many beautiful days for the bees.

December: Bees will continue to fly on Warmth and Light days. The hive entrances need to be checked regularly to prevent mice entering.

An example of a crop rotation for a domestic garden

Plot 1 receives manure compost in the autumn. A few corners should left for sowing radishes.

Plot 2 receives plant compost in the autumn.

Plot 3 is manured in the autumn partly with composted animal manure and partly with plant compost.

Plot 4 receives composted animal manure in the autumn.

Plot 5 is given manure compost shortly before planting.

Possible planting scheme in the first year:

Plot 1	*Plot 2*	*Plot 3*	*Plot 4*	*Plot5*
White cabbage	Carrot	Peas	Potatoes	Strawberries
Red cabbage	Parsnip	Broad beans	Potatoes	
Savoy	Salsfy	Sugar peas	Potatoes	
Cauliflower	Beetroot	Dwarf beans	Potatoes	
Broccoli	Onion	Pole beans	Potatoes	
Curly kale	Celery	Sweet corn	Jerusalem artichokes	
Brussels sprouts	Leek	Paprika	Jerusalem artichokes	
Kohlrabi	Leaf beet	Cucumber	Jerusalem artichokes	
Radish	Fennel	Courgette	Jerusalem artichokes	
Winter radish	Parsley	Tomato	Jerusalem artichokes	

Plant different species so you can hoe them separately

Salads and cabbages can be grown together

Plot 3 is only manured for the last four crops. The first five are legumes and do not like manure.

In the following years the plots are changed around. The plants of plot 1 move to plot 2 and those on plot 2 to plot 3, etc. Spinach, lettuce, corn salad, orach and endives may be planted as catch crops before or after the main crop. They are not so dependent on the crop rotation but should not be grown on the same ground in the same year. The areas reserved for summer and winter radishes receive no manure. It is important to ensure that all the plants of plot 1 remain on one plot and are moved together from year to year. Flowers should only receive plant compost and for roses stinging nettle compost should be prepared.

A special trick for crop rotations under glass

It is often very difficult for market gardeners to create a meaningful crop rotation in greenhouses and polytunnels because of market pressure and the need to make it pay. So it happens that they are often forced to grow more cash crops than a healthy crop rotation allows. This is why some forty years ago, I made the suggestion of linking up with a farmer who could provide some straw bales. These are laid in the polytunnel and cut open. A soil and compost mix is added to the straw and cucumbers or tomatoes are planted in it. The preparations are used as they generally are in biodynamic systems. When three months later, the crops are harvested, the farmer gets his straw back in a condition eminently suitable for making compost and the gardener can start the cycle afresh.

Last year we compared the use of our own straw with that conventionally produced. We discovered that the daily watering of the brought in straw resulted in enormous fungal growth and as a result of which the cucumbers grew sick. Using our own straw however they developed strongly and cropped well. We intend to repeat this trial in the coming year and use straw from the four main types of grain, for there are surely qualitative differencies between them that affect the cucumber and tomato crops.

Straw bale with compost

Compost made from special materials

Following the recommendation of a park warden in Baden Baden, we prepared a special rose compost made from stinging nettles. The nettles are mowed in May when they are growing strongly. Then as the pile is being built small quantities of ripe compost are added, cow pat (barrel) preparation is sprayed, the compost preparations are inserted and everything is covered with freshly cut lawn mowings. If there is very little rain it will need watering occasionally. By the autumn it will have become a wonderful, black and earthy compost ready for spreading on the rose beds in October. While doing this it is important to remember to add some to the rhubarb bed. It will respond well and in the next year produce strong red tinged stems with a spicy aroma.

Raspberry cultivation

Again and again we receive letters from readers of the *Sowing and Planting Calendar* about the poor health of their raspberries. The raspberry is by nature a plant of the woodland edge. This means that it grows, not in full sunlight but in the semi shade and enjoys layers of leaf mould around its roots, perhaps mixed with pine needles. The finer varieties like to have compost when planted. A very good compost can be made of spruce branches. All the twigs used during November and December for dressing graves and in Christmas decorations, can be used for this purpose. Thicker branches should not be used. While the material is being heaped up small quantities of ripe compost should be added, some barrel preparation given and the heap moistened. If set up in spring and consciously cared for the compost will be ready in the autumn for spreading among the raspberries. It is also well suited for use in shrubberies. Raspberries will thrive under this treatment and fruit profusely without any cane blight. Flowers and herbs thus treated will also reward the gardener with abundant flowers.

Leaf covering of raspberries

Compost for flowering hedges

For our favourite flowering hedges which are of course woody plants, the gardener needs to prepare a special compost. They need compost produced from cellulose material to stimulate flowering. Bark chippings can be used for this. They should however not be used in the form in which they are purchased but need to go through a further composting process. A pile is made and for every cubic metre of bark chippings, 10 kg of chicken manure is added. Then the compost preparations are inserted and it is well watered and about 10–15 litres cow pat (barrel) preparation are poured on. The pile is then covered with leaves, straw or grass. It needs to be watered once or twice more in the following weeks. Applying this compost as a soil covering in the autumn will provide the flowering hedge plants with exactly what they need.

If fresh tree bark or wood shavings are used it is important to make sure that no wood treatment chemicals have been applied. This would be harmful to the hedge plants. Tree bark and wood shavings need composting for at least two years otherwise the click beetle will lay its eggs under the hedge and plague you with wireworm for years after.

For hedges, fir branches can be cut an composted

Agricultural crop rotations

This largely concerns arable farms. Many farmers begin their rotation with a two year clover grass ley. With repeated mowing a lot of organic matter is produced in the soil. Because a portion of the roots die back when the grass is cut, soil organisms are working continuously to create humus. This process is further encouraged if cow pat (or barrel) preparation is sprayed once in the evening after mowing, then a week later Horn Manure is sprayed (also in the evening) followed soon afterwards on a Leaf day with Horn Silica early in the morning. If the biodynamic preparations are not sprayed, stinging nettle tea can be used once in the evening after mowing followed on the next Leaf morning with yarrow flower tea. If this is not done there is a danger that the click beetle will lay eggs in the undecomposed cellulose substance of the soil and cause wireworm damage to the subsequent wheat crop. The larvae take several years to develop into adult beetles and during this time live on the crop roots. By applying the biodynamic sprays increased soil activity is stimulated and the best growing conditions are created for the following crop of wheat.

Wheat should be followed by a root crop. If potatoes, mangels or sugar beet are chosen and there is a flock of sheep on the farm, it is good to apply some composted sheep manure. Folding sheep before the winter ploughing has the same effect. It is good to apply cow pat (barrel) preparation when the manure is worked into the soil. If the root crops chosen are carrots and beetroots, another procedure should be followed. After the wheat, a crop of rye should be grown. It has deep roots and leaves behind a well mineralized soil. Carrots, onions and beetroots thrive in these condi-

Sheep manure on a pasture

tions. They benefit from the nitrogen which is brought into the soil with the air during the hoeing and cultivation work especially when carried out on Root days. According to research by Professor Ahrens, this nitrogen is rapidly incorporated into the humus. It is recommended that along with the hoeing and cultivating, Horn Silica preparation and occasionally also chamomile tea should be applied on Root days early in the morning. This will encourage growth and improve quality.

For roots that are to develop a high sugar content it is worth spraying Horn Silica in the autumn after 3.00 pm and when the Moon is in the Ram or Scales. This will help to ripen the proteins, lower the nitrate content and increase sugar. If hoeing is done with the Moon in the Goat there will be relatively few weeds.

After the root crops, the soil is ready for flowering plants such as rape, linseed and sunflowers. Rape enriches the soil with sulphur, important for a following cereal crop. Fodder mallow which activates phosphorous for a succeeding cereal can also be considered. Peas, beans and lentils can be planted too for harvesting both green and dried. There is a good market for them among green consumers.

The manures from different farm animals bring special qualities that can be used to benefit specific crops. Oats and maize react particularly well to horse manure compost. If in addition Fruit days are chosen to spray Horn Silica early in the morning as well as perhaps yarrow flower tea, the size of the grains and level of yield will be noteworthy. These oats will be ideal for producing oat flakes.

Horse manure

Undersowings

In relation to cereal growing I would now like to refer to the practice of undersowing. On my parents' farm red clover was sown immediately after the oats were planted. In some years the clover grew so strongly that harvesting the oats was difficult and the clover had to be shaken from every forkful before binding. It then had to be fed to the cows taking care to sprinkle it with sufficient chaff so as to prevent bloating.

The undersowing of spring barley with lentils was more successful. The lentils climbed up the barley stems, the seeds ripened simultaneously and after the harvest they could be easily separated. Lentils planted as a single crop are difficult to harvest in a wet year because they lodge easily. We have since learnt the value of undersowing rye and wheat with cornflowers, corn cockle and phacelia. These all helped to increase phosphorous availability, improved grain yields by around 12% to15% and left more available phosphorous in the soil after harvest than had been there before.

Rye with cornflowers

The strawberry

If we observe how the wild strawberry develops through the year we will note that it begins to flower very early in spring. The leaves are then very small. One has the impression that the flowers belong to the leaves of the previous year. The flower stalks gradually grow and become quite long. The plant develops fruit and extends itself beyond the neighbouring plants in order to meet the sunlight. The seeds set on the outside of the fruits and only when they are ripe do the leaves start to grow and then become very large. Sheltered by these large leaves and close to the roots, the flower buds for the following year begin to develop.

Unfortunately, with the many cultivated varieties very little account is taken of this natural process of development. First of all the soil is heavily manured before planting and this forces the plants into strong leaf growth. Then the flowers develop on short stalks and heavy fruit are produced, so heavy that the stalks can't carry them. Afterwards the fruits lie so close to the soil that not only the flavour but the health of the whole fruit is affected.

If we consider the strawberry correctly in its annual cycle, the year begins after

Strawberries

Wild strawberries

the picking season. The plant appears to want to run away from its place and sends its children to wander off on runners far from the mother plant. Some of these runners can be taken and planted elsewhere using a small amount of earthy compost. Others can be retained close to the mother. Ripe compost is then added between the rows, worked lightly into the soil and the ground is sprayed with Horn Manure preparation. We soon notice the leaves beginning to grow. It is now that the flower buds for next year are being formed. At this point the plant needs a Light impulse. This is provided by the Horn Silica, a preparation which is made from rock crystals. It is stirred for an hour at dawn on a Fruit day and sprayed as a fine mist over the leaves. This provides the plants with a Light and Warmth impulse and influences the dormant buds which lie patiently waiting for the next year.

This etheric seed as it may pictorially be described, now needs a ripening experience. To provide this, the Silica preparation is given a second time on a Fruit day but this time between 3.00 and 4.00 pm. The plant is now able to make this Warmth impulse fully its own. It takes on a red glow as an outer picture for this process, before withdrawing into itself and waiting like a seed until spring when it will put out some relatively small leaves yet have stalks strong enough to carry the flowers and fruit.

The leaves of the previous year lie wilting on the ground while new growth is developing. None of next year's fruit will lie directly on the earth to absorb fungal spores. Instead everyone will be able to enjoy its magical scent and wonderful flavour. What happens in the plant if the leaves are simply mown down in the autumn and there is no possibility for this ripening to take place?

Another way of planting new strawberry beds

Strawberries are left to grow in the same bed for several years. New ones are prepared in August. They are cared for in the way described above. We use a distance of 1.2 metres between the rows. For the first year this is too wide. The whole bed is given compost, not only where the strawberries are to be planted. The large space between the rows is initially used for other plants and because strawberries are only cultivated on Fruit days, accompanying plants must also be suited to these days. Fruit plants such as small varieties of tomato, peppers, courgettes, maize etc. can be used for inter-planting. However, the area must not be over-planted.

A strawberry plantation interpalnted with spring onions which are also cultivated on Fruit days

In the second year, after the strawberries have been harvested, the runners are laid back along the rows and the areas between them are forked over and sprayed with Barrel Preparation. Flowering plants that enrich the soil with their root excretions such as cornflowers, phacelia or mustard, can be sown or planted in these areas. This will benefit the strawberries in the next year. In the third year it is usually possible to sow dwarf beans or lupins between the rows to supply some extra nitrogen. By the fourth year the strawberry rows will have so extended themselves that only a tiny path will remain between them. This method has produced a rich harvest of fruits of fine aroma and flavour that we have been able to offer to our guests throughout the year. After the old bed is removed the soil is ideal for growing cabbages.

If you keep hens, compost the droppings separately to get a good poultry manure

Poultry manure

Poultry manure is often offered for sale to domestic gardeners. It should however never be used fresh but always made into compost together with plant material. To do this it must be kept moist, otherwise it will heat up and its best nutrients evaporate. It is unsuitable for leaf plants but may be used on all fruit plants and flowers. If it is well composted with plenty of plant material, it can be applied to good effect on fruit bushes and trees in the autumn. It is most effective when worked lightly into the soil.

The fertilizing effect of animal manures

We have often pointed out the special qualities of horse manure. It has a strong warming effect which leaf plants such as lettuce, spinach and corn salad are not keen on. These plants need to continue growing undisturbed in the leaf realm and this is not so possible with horse manure. (In referring to animal manure, I am of course speaking of composted manure). Horse manure tends to stimulate early seed formation.

This is precisely what we want avoid with leaf plants. They prefer cattle manure. The quiet unhurried assimilation of food, the multiphase digestive process of the ruminant, are all qualities that produce a manure ideally suited to the leafy part of the plant. It should however not be used in a fresh state but must first go through a composting process to become earthy material. This is the finest substance that leaf plants including brassicas could ever wish for.

Now consider the pig. It eats everything. It has a relatively simple digestive system. If the manure is piled up for composting, it only warms up slightly and it takes a long time for fungi to establish themselves. Trials show that celery and parsnips do well with pig manure.

Sheep do not like being kept in paddocks. They enjoy eating the weeds and herbs of the grass verges. No sooner do they start feeding than their peristaltic movements begin and the first droppings appear. If sheep manure is set up for composting it must be kept very moist otherwise it heats up and its best qualities will be lost. It is good for all seed and fruiting crops. Comparative trials indicate that potatoes react well. They produce good yields, show healthy growth and have a special quality.

We kept mountain goats for a number of years. When these animals are led out to graze, they behave in a very different way to sheep. Pear tree branches in a neighbour's garden, or indeed prize roses were devoured in a split second while grasses and herbs growing by the wayside proved uninteresting. Near hedges or woodland trees they would stand on their hind legs and munch leaves or needles to their heart's content. Their manure behaves in a similar way to that of sheep.

Compost with chopped fir branches, horse and cattle manure

The care of top fruit, vines and soft fruit plantations

Responding to questions from various orchardists and wine growers we would like to offer the following recommendations regarding the care of all top fruit, vines and soft fruit plantations with the exception of bilberries:

In the northern hemisphere the orchard year begins at the beginning of November when about 1 tonne/ha of well rotted compost should be spread. Three successive soil cultivations should be undertaken on Fruit days once the leaves have fallen. At the same time the barrel preparation should be applied as a spray (along with the tree paste) to the trunks and branches. This is an alternative to the previously recommended tree painting.

To make the tree paste: Place 10 kg of cow pats into a wooden container and add 10 kg clay, 2.5 kg Basalt meal, 2.5 kg wood ash and 1 litre of whey. Everything should then be mixed together and moved about thoroughly with a spade for one whole hour (dynamized). The paste can be kept in a dark and cool place and used over several weeks.

When stirring the barrel preparation 1 kg paste should be added to 100 litres of water and stirred for 20 minutes. Then after allowing it to settle, the liquid is sieved through a pair of nylon stockings to prevent the sprayer from blocking. It should be sprayed out three times on Fruit days while cultivating the soil.

On Fruit days during March, Horn Manure preparation should also be sprayed three times during soil cultivation and at the same time be applied to the trunks and

Red currant

Black currant

Gooseberries

branches. In spring choosing a Leaf day when the first leaves have opened, a spray of stinging nettle tea should be given towards evening. A first spraying of Horn Silica can be applied to the newly emerging leaves when blossom time is over on a Fruit day. Stirring should begin half an hour before sunrise and be sprayed out immediately.

Early in the morning on Flower days dandelion flower and chamomile tea can be applied as a spray. Yarrow flower tea can be sprayed on Fruit days independently from the Horn Silica.

The Horn Silica and teas sprays are always applied to the leaves. To support bud formation in the right way Horn Silica may be sprayed on the following dates in 2005: July 1, 10, 11, 12 until 7.00 and on July 20, 21, 28 and 29 July. If summer pruning is necessary select Flower and Fruit days from July 19 July until the end of the month or in later areas from July 15 until 28.

To harvest fruit for winter storage choose Flower or Fruit days during the period when the Moon is ascending — that is not during planting time. After harvest but before leaf-fall Horn Silica should be sprayed on Fruit days in the afternoon from 3.00 pm onwards. Spraying at this time helps to ripen the wood and increase resistance to fungus attack in the following year.

Insect plagues

It is always a good idea for gardeners consciously to observe the world of plants and animals and their relationships to cosmic rhythms. In 2004 we became aware once again of how the influences coming from the constellation of the Bull via the Sun, Moon and planets affect such insects as the Colorado Beetle, Rose Chafer and Garden Beetle. These are referred to as chitinous creatures. If their reproductive power is very strong it is possible, by burning a very few specimens, to regulate their numbers and prevent them taking on plague proportions. For details, see *Animal and insect pests,* page 42.

Top: Colorado Beetle,
Leptinotarsa decemlineata

Centre: Rose Chafer
Cetonia aurata

Bottom: Garden Beetle
Phyllapertha horticola

Further Reading

Biodynamics. New Zealand Bio-Dynamic Association. Random House, NZ.
Bockemühl, Joachim, *In Partnership with Nature*. Anthroposophic Press, USA.
Castelliz, Katherine, *Life to the Land: Guidelines to Biodynamic Husbandry,* Lanthorn Press.
Cloos, Walther, *The Living Earth*, Lanthorn Press.
Colquhoun, Margaret and Axel Ewald, *New Eyes for Plants*, Hawthorn Press.
Conford, Philip, *The Origins of the Organic Movement*, Floris Books.
Davidson, Norman, *Sky Phenomena*, Anthroposophic Press, USA and Floris Books.
Grohman, Gerbert, *The Plant*, 2 volumes, Biodynamic Literature, USA.
Grotzke, Heinz, *Biodynamic Greenhouse Management,* Biodynamic Literature, USA.
Koepf, H.H. *The Biodynamic Farm*. Anthroposophic Press, USA.
—, *Research in Biodynamic Agriculture*, Biodynamic Farming & Gardening Association, USA.
Kranich, Ernst Michael, *Planetary Influences upon Plants*, Biodynamic Literature, USA.
Moore, Hilmar, *Rudolf Steiner's Contribution to the History and Practice of Agricultural Education*, Biodynamic Literature, Rhode USA.
Pfeiffer, Ehrenfried, *Soil Fertility*, Lanthorn Press.
—, *The Earth's Face*, Lanthorn Press.
—, *Weeds and What they Tell*, Biodynamic Farming and Gardening Association, USA.
Philbrick, John and Helen, *Gardening for Health and Nutrition*, Garber Books, USA.
Podolinski, Alex, *Biodynamic Lectures*, 2 vols. Gavemer Foundation, Australia.
Remer, N. *Laws of Life in Agriculture*, Biodynamic Farming & Gardening Association, USA.
—, *Organic Manure*, Biodynamic Farming & Gardening Association, USA.
Sattler, F. & E. von Wistinghausen, *Biodynamic Farming Practice*, Biodynamic Agricultural Association.
Schilthuis, Willy, *Biodynamic Agriculture*, Floris Books.
Schultz, Joachim, *Movement and Rhythms of the Stars*, Floris Books, Edinburgh.
Soper, John, *Biodynamic Gardening*, Biodynamic Agricultural Association.
Steiner, Rudolf, *Agriculture (A Course of Eight Lectures)*, Biodynamic Literature, USA.
Storl, Wolf, *Culture and Horticulture*, Biodynamic Literature, USA.
Thun, Maria, *Gardening for Life*, Hawthorn Press.
—, *Results from the Biodynamic Sowing and Planting Calendar* (formerly *Work on the Land and the Constellations*) Floris Books, Edinburgh.
von Keyserlink, Adelbert Count, *The Birth of a New Agriculture*, Temple Lodge Publishing.
—, *Developing Biodynamic Agriculture*, Temple Lodge Publishing.

Biodynamic Associations

Australia: Biodynamic Agricultural Association, PO Box 54, Bellingen, NSW 2454. *Tel:* 02 6655 0566. *Fax:* 02 6655 0565. *Email:* bdoffice@biodynamics.net.au *Web:* www.biodynamics.net.au

Canada: Demeter Canada, 115 Des Myriques, Catevale Que. J0B 1W0. *Tel:* 819-843-8488. *Email:* laurier.chabot@sympatico.ca *Web:* www.demetercanada.com

Ireland: Biodynamic Agricultural Association, The Watergarden, Thomastown, Co. Kilkenny. Tel/*Fax:* 056-54214. *Email:* bdaai@indigo.ie *Web:* www.demeter.ie

New Zealand: Biodynamic Farming & Gardening Association, PO Box 39045, Wellington Mail Centre. *Tel:* 04-589 5366. *Fax:* 04-589 5365. *Email:* biodynamics@clear.net.nz *Web:*www.biodynamic.org.nz

South Africa: Biodynamic and Organic Agricultural Association, PO Box 115, 2056 Paulshof. *Tel:* 011-803 1688 *Fax:* 011-803 7191.

UK: Biodynamic Agricultural Association (BDAA), Painswick Inn, Gloucester Street, Stroud GL5 1QG. Tel/*Fax:*01453 759501. *Email:* office@biodynamic.org.uk *Web:* www.biodynamic.org.uk

USA: Biodynamic Farming and Gardening Association, 25844 Butler Road, Junction City, OR 97448. *Tel:* 888-516-7797 or 541-998-0105. *Fax:* 541-998-0106. *Email:* biodynamic@aol.com *Web:*www.biodynamics.com